Are Generational Categories Meaningful Distinctions for Workforce Management?

Committee on the Consideration of Generational Issues
in Workforce Management and Employment Practices

Board on Behavioral, Cognitive, and Sensory Sciences

Division of Behavioral and Social Sciences and Education

A Consensus Study Report of

The National Academies of
SCIENCES · ENGINEERING · MEDICINE

THE NATIONAL ACADEMIES PRESS
Washington, DC
www.nap.edu

THE NATIONAL ACADEMIES PRESS   500 Fifth Street, NW   Washington, DC 20001

This activity was sponsored by the U.S. Army Research Institute for the Behavioral and Social Sciences (ARI) and was accomplished under Grant Number W911NF-19-1-0012. The views and conclusions contained in this publication are those of the authors and should not be interpreted as representing the official policies, position, or decision, either expressed or implied, of the ARI or the U.S. Government, unless so designated by other documents. The U.S. Government is authorized to reproduce and distribute reprints for Government purposes notwithstanding any copyright notation herein. Support for the work of the Board on Behavioral, Cognitive, and Sensory Sciences is provided primarily by a grant from the National Science Foundation (Award No. BCS-1729167). Any opinions, findings, conclusions, or recommendations expressed in this publication do not necessarily reflect the views of any organization or agency that provided support for the project.

International Standard Book Number-13: 978-0-309-67732-5
International Standard Book Number-10: 0-309-67732-7
Digital Object Identifier: https://doi.org/10.17226/25796
Library of Congress Control Number: 2020944331

Additional copies of this publication are available from the National Academies Press, 500 Fifth Street, NW, Keck 360, Washington, DC 20001; (800) 624-6242 or (202) 334-3313; http://www.nap.edu.

Suggested citation: National Academies of Sciences, Engineering, and Medicine. 2020. *Are Generational Categories Meaningful Distinctions for Workforce Management?* Washington, DC: The National Academies Press. https://doi.org/10.17226/25796.

The National Academies of
SCIENCES · ENGINEERING · MEDICINE

The **National Academy of Sciences** was established in 1863 by an Act of Congress, signed by President Lincoln, as a private, nongovernmental institution to advise the nation on issues related to science and technology. Members are elected by their peers for outstanding contributions to research. Dr. Marcia McNutt is president.

The **National Academy of Engineering** was established in 1964 under the charter of the National Academy of Sciences to bring the practices of engineering to advising the nation. Members are elected by their peers for extraordinary contributions to engineering. Dr. John L. Anderson is president.

The **National Academy of Medicine** (formerly the Institute of Medicine) was established in 1970 under the charter of the National Academy of Sciences to advise the nation on medical and health issues. Members are elected by their peers for distinguished contributions to medicine and health. Dr. Victor J. Dzau is president.

The three Academies work together as the **National Academies of Sciences, Engineering, and Medicine** to provide independent, objective analysis and advice to the nation and conduct other activities to solve complex problems and inform public policy decisions. The National Academies also encourage education and research, recognize outstanding contributions to knowledge, and increase public understanding in matters of science, engineering, and medicine.

Learn more about the National Academies of Sciences, Engineering, and Medicine at **www.nationalacademies.org**.

The National Academies of
SCIENCES · ENGINEERING · MEDICINE

Consensus Study Reports published by the National Academies of Sciences, Engineering, and Medicine document the evidence-based consensus on the study's statement of task by an authoring committee of experts. Reports typically include findings, conclusions, and recommendations based on information gathered by the committee and the committee's deliberations. Each report has been subjected to a rigorous and independent peer-review process and it represents the position of the National Academies on the statement of task.

Proceedings published by the National Academies of Sciences, Engineering, and Medicine chronicle the presentations and discussions at a workshop, symposium, or other event convened by the National Academies. The statements and opinions contained in proceedings are those of the participants and are not endorsed by other participants, the planning committee, or the National Academies.

For information about other products and activities of the National Academies, please visit www.nationalacademies.org/about/whatwedo.

Acknowledgments

On behalf of the Committee on the Consideration of Generational Issues in Workforce Management and Employment Practices, we thank the many people who contributed their time and expertise to assist in the committee's work and the preparation of this report. The study was initiated by the U.S. Army Research Institute for the Behavioral and Social Sciences (ARI), and we are particularly grateful for the guidance and support provided by Gerald (Jay) Goodwin, ARI, to facilitate the committee's work.

As discussed in this report, the nature of work and the composition of workers are changing rapidly in the 21st century. Although there have been many other events, the COVID-19 pandemic that emerged during the final stages of this study has demonstrated just how rapidly things can change and emphasized to organizations the importance of having the capabilities to adjust their workforce policies to new environments. Regardless of whether change is gradual and incremental or rapid and catastrophic, employers are often challenged to find new ways of managing their workforces across the employment life cycle. The U.S. military is not immune to these changes and is faced with many of the same challenges as other employers in both the private and public sectors.

Advice on managing multiple generations in the workforce is quite prevalent but sometimes contradictory. The committee was tasked to review the state and rigor of the empirical work related to generations and assess whether generational categories are meaningful in tackling workforce management problems.

To fulfill its charge, the committee heard from numerous people, including researchers, human resources professionals, military personnel officers, and corporate speakers. We are grateful to the input provided by these experts during our meetings and workshops and would like to thank the following presenters: Alexander Alonso, Society of Human Resource Management; David Autor, Massachusetts Institute of Technology; Peter Cappelli, Wharton Business School; Brian Carter, The Brian Carter Group; David Chu, Institute for Defense Analyses; Philip Cohen, University of Maryland; David Costanza, The George Washington University; Jennifer J. Deal, Center for Creative Leadership; Eric Dunleavy, DCI Consulting, director, Personnel Selection and Litigation Support Division; Richard Fry, Pew Research Center; Curtis L. Gilroy, retired, director, accession policy, Office of the Under Secretary for Personnel and Readiness, U.S. Department of Defense (DoD); Rick Guzzo, Mercer; Lernes "Bear" Hebert, acting deputy assistant secretary of defense for military personnel policy, Office of the Under Secretary for Personnel and Readiness, DoD; Steve Henderson, Bureau of Labor Statistics; Martha Hennen, Securities and Exchange Commission; Kim Lear, Inlay Insights, Inc.; Don Lustenberger, DCI Consulting; Sean Lyons, University of Guelph; Haig Nalbantian, Mercer; Frederick Oswald, Rice University; Cort Rudolph, Saint Louis University; Dietram Scheufele, University of Wisconsin–Madison; Christine Selph, Deloitte; William J. Strickland, retired CEO, Human Resources Research Organization (HumRRO), and retired colonel, U.S. Air Force; Jean M. Twenge, San Diego State University; Stephen E. Watson, Human Capital Management, U.S. Navy; Cortney Weinbaum, RAND; and Ken Willner, Employment Law Department, Paul Hastings, LLP.

This report, the result of the committee's work, contains our final conclusions and recommendations. Its preparation would not have been possible without the contributions and hard work of many individuals. The members of the committee dedicated their time and energy to collecting information, discussing alternative approaches, and drafting the report. The National Academies staff facilitated all aspects of the committee's work. Special thanks go to Julie Schuck, the study director, who performed a tremendous amount of research for the committee and kept the study organized and moving forward; Erin Hammers Forstag, science writer, who assisted with editing and drafting text for the report; Jeanne Rivard and Tina Winters, who provided research support in managing the generational literature and report elements; and Jacqueline Cole, Thelma Cox, and Anthony Mann, who handled the logistics for the committee and our invited guests at various stages of the project. Barbara Wanchisen, director of the Board on Behavioral, Cognitive, and Sensory Sciences, and Adrienne Stith Butler, associate board director, provided guidance to the committee throughout the study. In addition, the committee acknowledges the contributions of the Academies Research Center and Rebecca Morgan in searching databases and assembling the generational literature for this study and other staff members Kirsten Sampson Snyder, director of reports for the Division on Behavioral and Social Sciences and Education (DBASSE), and Monica Feit, DBASSE deputy executive director, who provided guidance and facilitated the report review process.

This Consensus Study Report was reviewed in draft form by individuals chosen for their diverse perspectives and technical expertise. The purpose of this independent review is to provide candid and critical comments that will assist the National Academies of Sciences, Engineering, and Medicine in making its published report as sound as possible and to ensure that the report meets institutional standards for quality, objectivity, evidence, and responsiveness to the study charge. The review comments and draft manuscript remain confidential to protect the integrity of the deliberative process.

We thank the following individuals for their review of this report: Peter Cappelli, Center for Human Resources, Wharton Business School, University of Pennsylvania; G. Marius Clore, Laboratory of Chemical Physics, National Institutes of Health; Jose M. Cortina, Department of Management and Entrepreneurship, School of Business, Virginia Commonwealth University; Lisa Finkelstein, Department of Psychology, Northern Illinois University; Emma Parry, Human Resource Management, Cranfield School of Management; Paul R. Sackett, Department of Psychology, University of Minnesota; and Donald M. Truxillo, Department of Psychology emeritus, Portland State University.

Although the reviewers listed above provided many constructive comments and suggestions, they were not asked to endorse the report's conclusions or recommendations, nor did they see the final draft of the report before its release. The review of this report was overseen by Howard M. Weiss, School of Psychology, Georgia Institute of Technology and Jonathan S. Skinner, Department of Economics, Dartmouth College. They were responsible for making certain that an independent examination of this report was carried out in accordance with the standards of the National Academies and that all review comments were carefully considered. Responsibility for the final content rests entirely with the authoring committee and the National Academies.

Nancy T. Tippins, *Chair*
Julie Anne Schuck, *Study Director*

Contents

Summary

eadlines frequently appear that purport to highlight the differences among workers of different generations and explain how employers can manage the wants and needs of each generation. But is each new generation really that different from previous ones? Are there fundamental differences among generations that impact how they act and interact in the workplace? Or are the perceived differences among generations simply an indicator of age-related differences between older and younger workers or a reflection of all people adapting to a changing workplace? To answer these questions, the National Academies of Sciences, Engineering, and Medicine was asked by the U.S. Army Research Institute for the Behavioral and Social Sciences to appoint an expert committee to review the scientific literature regarding generations in the workforce.

The Committee on the Consideration of Generational Issues in Workforce Management and Employment Practices included experts in management, industrial and organizational psychology, sociology, economics, research methods and statistics, learning sciences, adult development, personality and psychology, discrimination and diversity, and military personnel. The committee was tasked to assess the scientific literature concerning generational attitudes and behaviors in the workforce, to reach consensus on the state of this research, and to evaluate whether the concept of generations promotes understanding of the workforce and facilitates its management. The committee was also asked to make recommendations for directions for future research and improvements to employment practices.

The committee examined current workforce challenges in several job sectors in the United States. It collected hundreds of articles in the scientific literature on the topic of generations in the workforce, considered some of the multitude of pieces in the popular press on the same topic, and weighed other research on work and human capital. There is debate on whether there are more similarities than differences across generations of workers and whether generational categories are meaningful groupings in which to distinguish workers. The term "generation" has been used in many ways and has a range of definitions, but is often used to identify a group of people by their birth years.

As discussed in this report, many of the findings comparing different generations of workers are based on data collected at one point in time. In this case, observed differences cannot be tied to specific generational characteristics with certainty because they also can be due to age-related differences.

But what about the notion that today's young workers are different from young workers years ago? Some research has compared work values among young adults over time. Still, observed differences cannot be tied to specific generational characteristics with certainty because they also can be due to period changes that have affected everyone in society.

So what does it matter if real or perceived differences among workers are labeled as age, period, or generational differences? Given the existing hype on generations in the workforce in the popular discourse, it is important for research to attempt to distinguish generation effects from age and period effects. Not doing so limits the utility of research findings to inform management decisions. The best research would enable employers to consider, for example, whether the characteristics they observe in young recruits and recent hires (1) will persist in this generation of workers, (2) will change as they age, or (3) are representative of societal changes more generally that are affecting all workers. Further, since birth year is a fixed characteristic of individuals, it is important to avoid stereotyping and labeling a group of workers with attributes that could change as they age or as shifts in the nature of work occur. These issues are discussed further below, beginning with a look at the changing nature of work, followed by a review of the generational literature and suggestions for improving research in this area, and concluding with guidance for workforce management.

A NEW WORLD OF WORK

Notable economic, military, and political forces and social adjustments have reshaped the organization of work in the United States. These changes include increasing globalization, rapid technological innovation, expansion of the service sector, deregulation, and shifts in employee–employer relationships. At the same time, the characteristics of the workforce have changed. The education levels and skills of workers have risen as more people have completed high school and sought college degrees. Growth in the employment rates of women and older workers, later retirements, and increasing racial and ethnic diversity in the U.S. population have all contributed to the demographic diversity of today's workforce. With this diversity comes a range of needs and expectations with respect to work and the workplace.

These broad societal changes have been accompanied by important changes in the social and technical context of work itself. There has been relatively large growth in high- and low-skill jobs and slower growth in middle-skill jobs, polarizing the workforce. High-skill jobs have become more complex, demanding greater creativity and adaptability to solve evolving rather than routine problems. The rise in nonstandard work arrangements—such as contracting—has complicated the relationship between workers and the organizations for which they work. With advancing technologies, many workers have more autonomy as to when, where, and how they conduct their work. At the same time, interdependence among jobs and team-based approaches to work have increased, making interpersonal skills of workers and communication strategies within organizations more important.

These broad and contextual changes have created a demand for new employment practices in many organizations. Employers are seeking guidance on how to develop effective policies and practices for recruitment and retention and how to best manage a diverse workforce in these new work environments.

GENERATIONS IN THE WORKPLACE

One of the key changes in the workplace, and the impetus for this study, has been an increase in the age diversity of the workforce. In the 1970s, 1980s, and 1990s, the vast majority of the workforce consisted of young workers, aged 16–34, and middle-aged workers, aged 35–54. Starting around 2000, the proportion of older workers,[1] aged 55 and up, started to rise. As of 2018, the proportion of older workers was nearing parity with the proportion of young and middle-aged workers, each representing about a third of the U.S. workforce. This increase in age diversity has generated much press about potential differences among generations in the workplace, with some authors claiming as many as five different generations in today's workplaces.

---

[1] Note that the age that defines "older" workers continues to be debated in the research literature. For the purposes of this report, labor force statistics in age groups from the U.S. Bureau of Labor Statistics are used, combined to present what might be three generations of workers.

The concept of generation has a long history of scholarly consideration. Early sociological theories proposed the idea of generational shifts to explain social progress. These theories considered how entire groups of people who were born around the same time in the same area could influence change in society, but they did not focus on describing or understanding individuals within a group. These early theories recognized the importance of historical events at salient human developmental stages but acknowledged that the impact of these events on individuals would vary.

More recently, popular ideas have emerged that have adapted the sociological theories in notable ways. These ideas emphasize the influence of significant historical events on individuals, and propose that these events lead to shared values and behaviors among individuals born between certain years. Individuals born during these years make up a generation (often considered a span of 20 years.) Some labels evolved to connect a significant event to a generation. For example, "baby boomers" were born during years of increased birth rates after World War II, while "millennials" were born in the 1980s and 1990s prior to the turn of the millennium.

In the wake of these popular ideas, researchers in psychology and business management who examine workforce and workplace issues have shifted the focus on generations away from the sociological perspective of understanding social change and toward an understanding of individual work-related values, attitudes, and behaviors. This new body of research uses generational terminology common in the popular press; for example, many researchers use such terms as "baby boomers," "generation X," and "millennials" to categorize people in their studies. Most of this research assumes that generations have a set of shared experiences and that these experiences shape the attitudes and values being measured. These experiences are largely undefined but often implied to be associated with significant events (e.g., the terrorist attacks of September 11, 2001) or social phenomena (e.g., the digital age[2]) that occurred during a group's formative years.

The committee found conceptual and methodological limitations with this new generational research. The literature has not taken an empirical approach to define sets of experiences or to investigate the mechanisms by which shared experiences would shape lasting attitudes and subsequent behaviors across a large group of people. Moreover, as discussed below, most studies of generational differences make no attempt to separate generation effects from age and period effects, making it difficult to draw strong conclusions about generational characteristics.

---

[2] The idea of growing up during the ubiquity of smartphones is a common argument for purported generational differences, but there has been no evidence that the prevalence of any new technology will change a specific cohort of people more so than it changes a society.

Conclusion 4-1: Many of the research findings that have been attributed to generational differences may actually reflect shifting characteristics of work more generally or variations among people as they age and gain experiences.

REVIEW OF THE GENERATIONAL RESEARCH

The committee was asked to review research on generational issues in the workforce, the body of which has been growing steadily over the past 20 years. The research generally has focused on two types of questions: (1) Are today's young workers different from today's older workers because of a generation effect? and (2) Are young workers now different from young workers in the past? These are not easy questions to answer scientifically. For the first question, it is difficult to separate generation effects from age effects; for the second, it is difficult to separate generation effects from period effects.

Age, Period, and Cohort Effects

The concepts of age, period, and cohort are foundational for understanding whether issues in the workforce can be attributed to generational differences. Generations often are defined simply by their birth years, and generational researchers usually combine multiple birth years to define a cohort of people. When researchers look for generational differences, they need to be rigorous in their approach to distinguish cohort effects from age and period effects:

- **Age effects** are considered developmental influences resulting from biological factors or maturation that occur in all people. For example, age-related changes in muscle fibers create differences in physical strength, on average, between younger and older workers.
- **Period effects** are considered social influences that affect everyone in society. For example, while young adults today are more likely to have a cellphone than young adults 20 years ago, it also is true that all adults are more likely to have a cellphone today than was the case 20 years ago as a result of technological and societal shifts.
- **Cohort (or generation) effects** are considered social influences that predominantly affect only a certain group of people who share a defining characteristic. For example, cohorts have been defined by birth year, graduation year, or a shared experience such as working in the automotive industry during a particular period. A cohort effect, for example, might be observed in African Americans who were adolescents in the 1950s and 1960s. This cohort is likely to

have had much different experiences from those of other groups of people in the United States during that period, which could have shaped lasting differences between them and other groups.

The distinction between period and cohort effects can be difficult to appreciate, but it can begin to be statistically demonstrated with the right set of data on individuals over time. Events around the COVID-19 pandemic provide an interesting case in which both period and cohort effects may play out over time. For example, a period effect would be a shift in certain behaviors or attitudes (e.g., increased anxiety about health or job security) among all people, regardless of age, as a result of experiences during the pandemic. A cohort effect, on the other hand, would be changes in attitudes, behaviors, or outcomes for a limited group of people. For example, projected cohort effects resulting from COVID-19 experiences may be observed in groups defined by job sector, health condition, or socioeconomic status. The effects of COVID-19 on those who work in jobs in the service sector that require close personal contact are likely to be different from the effects on those with office jobs, who can adjust more easily to shifting levels of remote and virtual work. Cohort effects by generation that are significantly larger than general period effects are unlikely. While many young adults trying to enter the workforce during the pandemic face challenges and may have to weather long-term impacts on their careers and earnings, these consequences are likely to vary by type of occupation.

Limitations of Research Designs

The research designs used for generational research vary in their sophistication and their limitations. The vast majority of studies reviewed by the committee applied cross-sectional (i.e., single time point) designs to convenience samples.[3] Some studies used cross-temporal meta-analyses, and other studies used qualitative methods. None of these methods can separate generation effects from age and period effects. Only a few studies used complex multilevel statistical models applied to nested datasets (i.e., data

---

[3] In the context of this report, cross-sectional designs refer to methods of comparing people of different ages using an instrument (e.g., a survey) administered to a single sample at a single point in time; cross-temporal meta-analyses refer to a statistical approach of combining results from studies conducted at different points in time, usually with samples of a similar age (e.g., regularly administered surveys of high school seniors); and multilevel models applied to nested datasets, or age, period, and cohort (APC) models, refer to statistical methods that combine and analyze data on multiple individuals of different ages collected at different points in time (e.g., data from the General Social Survey) in an effort to separate out age, period, and cohort effects.

available from a series of studies or surveys conducted at different points in time) in order to separate the age, period, and cohort effects.

Cross-sectional analyses that use data from a single point in time to study workers of different ages (e.g., all workers in the year 2020) run the risk of confounding age and cohort effects. Because workers of different birth cohorts are also of different ages, observed differences could be due to age or cohort differences. In cross-temporal analyses that examine a single age group over time (e.g., 18- to 24-year-old workers in the 1980s, 1990s, and 2000s), cohort and period effects may be confounded. Because workers are being observed during different time periods, observed differences could be due to cohort or period differences. The qualitative studies reviewed also suffer from methodological limitations, including the use of purposive and convenience sampling, the risk of interpretation bias, and a failure to follow best practices in documenting data collection protocols and analysis processes. These shortcomings make it difficult to assess the value of the findings from qualitative studies.

These limitations weaken the internal validity[4] of research designs in answering the question of whether generational differences exist in the workforce because observed differences among groups may instead be due to age or period effects. Many studies also offer insufficient external validity[5] in that the findings are limited to a narrow setting and cannot be extended to all the members in a generation. The issue of *representativeness* relates to how well a sample reflects the population of interest. A convenience sample, which draws on only accessible members of a population (e.g., employees willing to fill out a survey), is not likely to be representative of generations. While a few studies have included steps to analyze data from nationally representative samples, the problem of generalizability remains largely unaddressed in the body of literature on generational attitudes in the workforce.

When these methods are used, researchers and users of the research need to understand the limitations of the methods and the available data and draw appropriate inferences from the findings.

> **Conclusion 4-2:** The body of research on generations and generational differences in the workforce has grown considerably in the past 20 years. Despite this growth, much of the literature suffers from a mismatch between a study's objectives and its research design and under-

---

[4] Internal validity refers to the trustworthiness of the research design and methods for selecting and engaging participants. It also reflects the extent to which a study makes it possible to eliminate alternative explanations for any findings.

[5] External validity refers to the extent to which findings from a study are generalizable and can apply to other settings.

lying data, which threatens both the internal and external validity of the work. The research designs and data sources rely too heavily on cross-sectional surveys and convenience samples, which limits the applicability and generalizability of findings.

Some researchers have employed research designs that apply multilevel models to nested datasets: statistical methods that combine and analyze data on multiple individuals of different ages collected at different points in time. These designs have significant advantages over other methods in distinguishing among age, period, and cohort effects. Such research has found little evidence for generational differences in work values. Rather, the evidence points to pronounced period effects, suggesting that changes are reflected in the workforce more broadly rather than in a specific generation of workers. However, since relatively few datasets with information relevant to workforce considerations are available for this type of analysis, the research questions that can be addressed are constrained.

Improving Future Research

Acknowledging the limitations of the existing generational literature on work-related attitudes and behaviors, the committee believes future research in this area will need some important modifications. Going forward, researchers should pay greater attention to their research designs and the questions that these designs can appropriately address.

Recommendation 4-1: Researchers interested in examining age-related, period-related, or cohort-related differences in workforce attitudes and behaviors should take steps to improve the rigor of their research designs and the interpretation of their findings. Such steps would include

- decreased use of cross-sectional designs with convenience samples;
- increased recognition of the fundamental challenges of separating age, period, and cohort effects;
- increased use of sophisticated approaches to separate age, period, and cohort effects while recognizing any constraints on the inferences that can be drawn from the results;
- greater attention to the use of samples that are representative of the target populations of interest;
- greater attention to the design of instruments (e.g., surveys) to ensure that the constructs of interest (i.e., measured attitudes and behaviors) have the same psychometric properties across time and age groups; and

- increased use of qualitative approaches with appropriate attention to documenting data collection protocols and analysis processes.

ALTERNATIVE PERSPECTIVES ON GENERATIONS FOR FUTURE RESEARCH

Inherent Appeal and Biases

Despite the fact that research has largely not produced evidence in support of generational differences, there is an inherent appeal to the notion that groups of people born at different times have certain attributes and values. Humans are inclined to categorize and generalize; these tendencies can be useful when deciding whether a situation is dangerous or simplifying a large amount of information. Social categorization of oneself and others—such as into generational categories—is a common manifestation of this process.

The notion of generations has become strongly socially constructed; that is, generational differences purportedly exist because they are frequently acknowledged in various contexts. In this sense, generational labels (e.g., baby boomers, millennials) have taken on a life of their own. Given their socially constructed nature, these labels can shape people's perceptions of themselves and other people, regardless of whether the underlying stereotypes are accurate.

While the concept of generations and the idea of generational differences can be useful in some instances, they can also lead to prejudice, bias, and stereotyping. People born in the same year or span of years may have some similar experiences, but they may also have very different experiences, depending on such factors as socioeconomic status, geographic location, education level, gender, and race/ethnicity. Some recent workplace research has shown that people's perceptions of generational stereotypes can influence how they perform and how they interact with others. Additional research is needed to fully understand the use and impact of generational stereotypes in the workplace.

Because generational beliefs and perceptions are not likely to reflect true attributes of members of any birth cohorts, they should be studied as generational stereotypes and biases. Areas ripe for research include examining how perceptions about generational qualities develop, what opportunities and challenges these perceptions present in the workplace, and what the implications are for organizations to address any prevalent misconceptions.

Multiple Influences on Worker Attributes

An additional task for researchers is to identify alternatives to the theory and research designs applied to date in the study of generational issues in the workforce. Future research should seek to examine the multiple influences that could be expected to affect similarities and differences among workers. The committee offers three perspectives for thinking about variations among workers: (1) lifespan development theories, (2) changes in the work context, and (3) the aging workforce. We recognize that further research may demonstrate other perspectives to be of value for understanding workforce issues. A lifespan development perspective considers the impact of historical events on human development while also stressing the importance of biological and cultural factors in explaining differences among people. This perspective differs from the traditional generational approach in acknowledging that people are influenced not only by broad historical events, but also by life events that are idiosyncratic to individuals.

A research perspective on changes in work context focuses on social and technical changes in the environments in which work takes place that occur as a function of broad social and economic adjustments. Research that takes context into account is useful for understanding how changes in work context drive different behavior patterns.

A research perspective on the aging workforce focuses on the emergent norms, practices, and behaviors that develop as a function of shifts in workforce demographics. Pointing to generational issues has masked real challenges in the management of a more age-diverse workforce. Neither generation nor age has been shown to be a reliable predictor of work-related outcomes. Research that considers job experiences and the cultural influences of an age-diverse workforce in addition to worker characteristics can be useful for understanding different behavior patterns.

Recommendation 5-1: Researchers interested in examining relationships between work-related values and attitudes and subsequent behaviors and interactions in the workplace should endeavor to identify and better understand alternative explanations for observed outcomes that supplement explanations associated with generations. This effort would include attention to generational stereotypes and biases that might exist among workers. Research should also seek to better understand the multiple factors that influence attributes of individual workers, including aging in the workplace, and the changes in the work context that affect the behaviors of all workers.

GUIDANCE FOR WORKFORCE MANAGEMENT

In the course of this study, the committee reviewed many documents addressing concerns about managing across generations in the workforce or managing a new generation of workers. Employers have asked what types of policies and practices will be effective for recruiting, retaining, and promoting job satisfaction for today's workers. Some discussions center around the idea that employers should take generational stereotypes into account when developing policies and practices. However, while dividing the workforce into generations may have appeal, doing so is not strongly supported by science and is not useful for workforce management. Research has shown there is much variation in worker needs and performance within all age groups.

> **Conclusion 6-1:** The notion of generational differences will continue to be appealing in the absence of compelling alternative explanations for real or perceived differences among people in the workplace. However, many of the stereotypes about generations result from imprecise use of the terminology in the popular literature and recent research, and thus cannot adequately inform workforce management decisions. Further, categorizing a group of workers by observable attributes can lead to overgeneralizations and improper assumptions about those workers, perhaps even discrimination.

Tailoring employment policies and practices to a specific group defined by birth year is unlikely to meet the needs of all members of that group, and may exclude members of another group for whom those policies and practices would be valuable. Moreover, when age, generational categories, or stereotypes about generations are used in the workplace to inform decisions or policies, the employer may be in violation of the Age Discrimination in Employment Act (ADEA) and various state and local laws on age discrimination.[6] Although these laws are based on age and do not explicitly address generational categories or stereotypes, a court could find that an employer who made a decision based on an employee's generation was using generation as a proxy for age. Employment decisions based on stereotypes about generations—such as refusing to put workers of a certain generation in a specific job position—could be particularly vulnerable to ADEA claims because Congress intended this act to combat pervasive stereotypes and stigmatization of older workers.

---

[6] As discussed further in report, the ADEA and some state laws apply only to workers 40 and over, while other state laws prohibit discriminating against workers of any age.

In its information gathering, the committee found that many employers struggle with recruiting and retaining talent. More specifically, we read about unfilled jobs in many sectors, notably the health care and service sectors, and shortfalls in recruitment targets for the military. There also have been challenges with turnover, including a rise in the number of employees eligible to retire, which have presumably led to the need for employers to reexamine their recruitment and retention strategies. Some evidence suggests that the changing nature of work is responsible for many of the concerns expressed by employers. Employers may need to revise their policies and practices in order to respond to these changes.

> **Recommendation 6-1:** In considering approaches to workforce management, employers and managers should focus on the needs of individual workers and the changing contexts of work in relation to job requirements instead of relying on generational stereotypes. Employers can be guided in making any needed changes to employment practices and policies by a thorough assessment of changes in their own work environment, job requirements, and human capital.

The goal of recruitment is to identify candidates whose preferences, skills, and abilities match the needs of the organization and the requirements of a specific job. People increasingly are entering and leaving the workforce at different life stages both for personal reasons and as a result of social and economic shifts in labor demands. Therefore, employers need to develop recruitment strategies that appeal to a range of people who are likely to be viable candidates. Indeed, many employers see the increasing diversity of the U.S. population as an opportunity to expand their recruitment pool and to match their workforce to their customer base. A diverse workforce can also have social and economic benefits for organizations.

Research has shown that an inclusive environment with attention to employee treatment and professional development reduces turnover. Steps taken to help employees feel safe, respected, and influential on the job and believe they have the ability to balance work and life needs can promote employee engagement with an organization. Further, the demand for continuous learning on the job has risen, driven in part by both broad and discrete changes to the organization of work—notably technological advances and hiring patterns that have led to institutional knowledge gaps between younger and older workers. Developing effective training programs requires attention to the needs of the organization and its employees, as well as the constraints within which the organization operates.

Training needs also have extended to the skills necessary to manage a range of workers with varying characteristics. While there are benefits to having a diverse workforce, there are also challenges entailed in addressing the needs of a range of workers and ensuring that this diversity produces the desired outcomes for organizations. There is no universal approach to increasing diversity and employee engagement; organizations have unique cultures requiring specific strategies that work in their particular context. The best advice and research evidence highlight the benefits of assessing one's own culture, engaging all levels of management in the assessment and the solutions thereby identified, and developing initiatives that go beyond procedural checklists to transform organizational culture as necessary.

The goal of effective workforce management is not to find permanent answers to recruitment and retention challenges. The nature of these challenges changes over time. As a result of the economic effects of the COVID-19 pandemic, for example, the recruiting challenges of January 2020 were substantially different from those just 4 months later, in May. Moreover, employees' needs and values change, and the missions of employers may adjust with broader societal changes. In addition, possible solutions are constantly evolving. For example, recently developed teleconferencing tools have enhanced the effectiveness of remote working and facilitated flexible work schedules and locations. Organizations must then evaluate the new policies and procedures they undertake to determine their impact on organizational effectiveness and the extent to which employees' needs are met. Thus, the committee recommends that organizations develop effective ways of regularly identifying changes in the work environment and employees' needs, determining available solutions to these problems, and evaluating those solutions.

Recommendation 6-2: Employers should have processes in place for considering and reevaluating on a regular basis an array of options for workforce management, such as policies for recruiting, training and development, diversity and inclusion, and retention. The best options will be consistent with the organization's mission, employees, customer base, and job requirements and will be flexible enough to adjust to different worker needs and work contexts as they change.

1

Introduction

Generational categories have become commonplace in the [human resources and business management publications] and have also been lent respectability by a growing academic interest in the subject.... Social scientific accounts are generally skeptical of the more sweeping uses of generations as units of analysis, but this has done little to temper other writing on the subject....The implication for management is that, if current recruits are qualitatively different from previous intakes, then perhaps it is employers who need to adapt to the new intake, rather than vice versa. (Williams, 2019, p. 2)

As practitioners have adopted the concept of generations, scholars have strived to examine the differences between generational groups and to provide evidence for the idea that these different groups have unique values, attitudes, preferences, and expectations both in and outside of the workplace. While many researchers are supportive of the concept of generations, a growing group of academics have questioned the validity of the idea that people are psychologically different according to when they were born. (Parry and Urwin, 2017, p. 140)

The past 20 years have seen significant discussions of generations in the workforce. These discussions can be found in myriad articles and books directed at personnel managers and human resources professionals focused on how to manage different generations in the workplace, as well as in increased research studies aimed at scientifically measuring and confirming the relevance of any differences among generations to work-related outcomes. Practitioners and scholars alike continue to debate whether such generational differences exist and whether generational categories are meaningful distinctions for workforce management.

At the same time, it is recognized that broad societal trends are affecting workers of all ages. Research in a number of disciplines has examined the impacts of social trends on work (Hoffman, Shoss, and Wegman, 2020), highlighting the changes and associated challenges faced by the modern workplace, including rapidly advancing technologies, an increasingly diverse workforce, trends in globalization, and new employer–employee relationships. Employers across various sectors, including the military, are attempting to recruit, manage, and retain workers while coping with these shifts, as well as new and evolving trends in worker preferences, such as improved work–life balance, flexible schedules, and later retirements.

THE COMMITTEE'S CHARGE

The Committee on the Consideration of Generational Issues in Workforce Management and Employment Practices was convened by the National Academies of Sciences, Engineering, and Medicine to examine the salience of generational categories to workforce issues. The study was sponsored by the U.S. Army Research Institute for the Behavioral and Social Sciences (ARI), whose mission is to maximize the performance and readiness of individuals and units within the Army through research on topics related to personnel performance and training.[1] The Army's interest in workforce issues is long-standing; the need to recruit, train, and retain a large number of personnel has led the Army to explore different ways of understanding potential recruits and current personnel.

The committee included experts in the areas of management, industrial and organizational psychology, research methods and statistics, learning sciences, adult development, personality and psychology, sociology, economics, discrimination and diversity, and military personnel. The committee was tasked with assessing the scientific literature on generational attitudes and behaviors in the workforce and evaluating whether the categorization of generations is a meaningful way of understanding and managing the workforce. The committee was also asked to recommend directions for future research and for any changes to employment practices. (See Box 1-1 for the committee's full statement of task.)

In undertaking its charge, the committee sought to identify and assemble the peer-reviewed literature on generational attitudes and behaviors relevant to the workplace, broadly defined, and to understand

---

[1] See https://www.consortium-research-fellows.org/work-sites/agencyid/3.

BOX 1-1
Statement of Task

An ad hoc committee will gather, review, and discuss the business management and the behavioral science literature on generational attitudes and behaviors in workforce management and employment practices. The committee will:

1. Evaluate theory, data and statistical methods used in order to make determinations on the rigor of the empirical work in this literature.
2. Assess whether generational categories (e.g., "boomers," "millennials") are meaningful distinctions vis a vis the workforce and its practices. Included issues will be recruitment, selection, assignment, training, learning, performance management, length of tenure in a job, and retention.
3. Provide conclusions and recommendations in terms of proposing a possible science agenda and/or changes that are warranted to better recruit and retain the best employees.

the common needs of employers and employees across many sectors, as well as the unique needs of the military. While the focus of this work was on assessment of the generational literature on work-related outcomes, the committee also drew on research in a range of fields, including economics, education, management, psychology, and sociology, to provide context for its assessment and advice for management, as well as identify future research needs.

This committee recognized that a previous National Academies study (National Research Council [NRC], 2002) examined the generational claims in the popular literature and found them to run counter to scientific findings (see Box 1-2). Since the publication of that letter report, much research has emerged in an effort to identify true generational differences, notably those related to work. This report examines the rigor of this recent research.

APPROACH TO THIS STUDY

During the course of this study, the committee held five meetings. These meetings consisted of a combination of information-gathering sessions open to the public and closed sessions in which the committee deliberated on this information and findings from its review of the relevant literature (see below), and developed conclusions and recommendations for this report. The committee also held two public workshops:

BOX 1-2
Findings from a Previous National Academies Study

In 2002, the Committee on Youth Population and Military Recruitment issued a letter report to Lieutenant General John A.Van Alstyne, deputy assistant secretary of defense for military personnel policy (NRC, 2002). This letter report, prepared as part of a larger 3-year study examining military recruitment challenges, trends in youth values, and the changing nature of work, was published in response to a request by the Office of Accession Policy to assess "the scientific quality of the popular literature characterizing various generations, with a particular focus on millennials" (NRC, 2002, p. 1).

The letter report focused on two claims in the popular literature: "[1] there are distinct generations with sharp differences among them, and [2] there are large and dramatic differences among youth cohorts in different generations" (NRC, 2002, p. 2). Its preparation was informed by the study committee's review of eight books (Copland, 1991; Howe and Strauss, 1993, 2000; Mitchell, 1995, 1998; Strauss and Howe, 1991, 1998; and Zemke, Raines, and Filipczak, 1999) that, while not considered part of the peer-reviewed scientific literature, were frequently cited in documents from the Department of Defense that had been examined as part of the larger study.

In its letter report, the committee concluded that these two claims ran counter to scientific findings. It argued that the notion of distinct generations with clear differences among them was not supported by research. Further, the committee reported that in its examination of trends in youth values, it found that existing data from longitudinal research showed much stability in attitudes and values among youth over time. Where change had occurred (e.g., change in seeing work as a central part of life), it had done so gradually, not sharply. The committee warned "against uncritical acceptance of claims for generational characteristics and [encouraged] careful examination of the scientific bases for any such claims" (NRC, 2002, p. 5).

- The first public workshop, "Trends in Workforce Management: Are Generational Labels Meaningful?," was held May 29, 2019, in Washington, DC. It explored recent and predicted societal and demographic trends in the United States with implications for the workplace, as well as how employers, with special attention to the military, have responded to these trends and to any evidence of generational differences in the workplace.
- The second public workshop, "Changes in the Work Environment: Societal Trends and Workforce Management," was held July 30, 2019, in Washington, DC. In this workshop, researchers from the fields of sociology, psychology, economics, and business management presented the evidence for changes in the workplace and the resulting challenges and opportunities for workforce management.

In addition, the committee heard from an expert on science communication to gain perspective on how it could be improved to strengthen connections between research and practice.

During the public sessions and workshops, the committee heard presentations from a number of stakeholders, including the sponsor, researchers, human resources professionals, military personnel officers, and corporate representatives (see Box 1-3).

The committee also was tasked to "gather, review, and discuss the business management and the behavioral science literature on generational attitudes and behaviors in workforce management and employment practices." In its literature search, the committee identified several recent literature reviews on this topic and more than 500 articles in the scientific literature published since 1980. (Appendix A details the committee's search strategy and describes the literature reviewed).

KEY CONCEPTS

This section explains how the concept of "generation" was used for this study and describes the foundational concepts of age, period, and cohort effects.

Generation

The scientific literature has defined the term "generation" in various ways. According to Merriam-Webster (2019), the term can refer to a number of concepts, including

- a body of living beings constituting a single step in the line of descent from an ancestor;
- a group of individuals born and living contemporaneously;
- a group of individuals having contemporaneously a status (such as that of students in a school) that each one holds only for a limited period; and
- the average span of time between the birth of parents and that of their offspring.

In the conduct of empirical research, a concept needs to be operationalized so it can be linked to variables that can be measured and studied. As discussed in Chapter 4, the concept of generation has been difficult to operationalize, and for many studies, birth cohort has been used as a proxy for generation. In this report, the committee has adopted a similar approach, using the term "generation" to denote a birth cohort, that is, a group of people born during a particular year or sequential set of years.

BOX 1-3
Invited Presenters at the Committee's
Public Sessions and Workshops

- Alexander Alonso, Society of Human Resource Management
- David Autor, Massachusetts Institute of Technology
- Peter Cappelli, Wharton Business School
- Brian Carter, The Brian Carter Group
- David Chu, Institute for Defense Analyses
- Philip Cohen, University of Maryland
- David Costanza, The George Washington University
- Jennifer J. Deal, Center for Creative Leadership
- Eric Dunleavy, Personnel Selection and Litigation Support Division, DCI Consulting
- Richard Fry, Pew Research Center
- Curtis L. Gilroy, Office of the Under Secretary for Personnel and Readiness, Department of Defense (retired)
- Gerald (Jay) Goodwin, U.S. Army Research Institute for the Behavioral and Social Sciences
- Rick Guzzo, Mercer
- Lernes "Bear" Hebert, Office of the Under Secretary for Personnel and Readiness, Department of Defense
- Steve Henderson, Bureau of Labor Statistics
- Martha Hennen, Securities and Exchange Commission
- Kim Lear, Inlay Insights, Inc.
- Don Lustenberger, DCI Consulting
- Sean Lyons, University of Guelph
- Haig Nalbantian, Mercer
- Frederick Oswald, Rice University
- Cort Rudolph, Saint Louis University
- Dietram Scheufele, University of Wisconsin-Madison
- Christine Selph, Deloitte
- William J. Strickland, Human Resources Research Organization (HumRRO) (retired) and Colonel, United States Air Force (retired)
- Jean M. Twenge, San Diego State University
- Stephen E. Watson, Human Capital Management, United States Navy
- Cortney Weinbaum, RAND
- Ken Willner, Employment Law Department, Paul Hastings, LLP

Age, Period, and Cohort Effects

Age is measured as time since birth and is a changing characteristic of individuals. An *age effect* occurs when individuals of different ages vary in the way they think, feel, and behave because of factors related to their stage of the life course. Age effects are considered developmental influences because they are a result of biological factors or maturation that occurs to all people regardless of when in history they were born and current historical conditions. For example, younger workers may be physically stronger (on average) than older workers because of age-related changes in muscle fibers.

Period is typically captured as the year of observation and is a changing characteristic of the broader sociohistorical context. A (time) *period effect* occurs when individuals change in the way they think, feel, and behave because of the events or social phenomena of a specific point in history. For example, the impact of a global pandemic might lead to increased anxiety for all people in a society at a given point in time, regardless of age group. After the pandemic had ended, everyone might express more apprehension about disease, even if at different levels, than they would have before the pandemic.

A cohort is a group of individuals with distinct characteristics or experiences. Cohorts are often defined as those individuals born in the same year and expected or known to have moved through their lives in concert and experienced major events at the same point in their development. The same idea applies to people who were born within a narrow set of birth years, which is why generational research often combines multiple birth years. Birth year is a fixed attribute of individuals. If a strong *cohort effect* is observed in statistical analysis, this would indicate, for example, that workers born in 1972 are categorically different from workers born in 1992 as a result of the differential influence of cultural, historical, and social events. A cohort effect differs from a period effect in that with a cohort effect, particular historical experiences influence a specific group of people because of their stage of development (or other unique characteristic) at the time of exposure, whereas a period effect impacts all people regardless of age. A cohort effect is unique to people born in a particular year or set of years because of when in their development they were exposed to particular events. For example, the events of an economic depression might make all people sensitive to financial losses after the depression (a period effect), or it might uniquely affect a group in their formative years (a cohort effect) because of the more negative emotional and economic impact on their earning potential at a time when they were entering or exploring the labor market.

For most studies of people and the variations among individuals over time, some aspects of age, period, and cohort all may contribute to the outcomes observed. The challenge for researchers is to identify which is the predominant influence. See further discussion of age, period, and cohort effects in Chapter 4.

ORGANIZATION OF THE REPORT

The remainder of this report consists of five chapters. Chapter 2 explores the macro economic, military, political, and social trends changing the nature of work and the workforce. Chapter 3 provides background on the origins of generational theory and its adaptions for use in the popular press, business advice, and research. Chapter 4 presents the committee's findings and conclusions from its review of the scientific research on generational attitudes and behaviors in the workforce and provides an overview of the conceptual and methodological issues that challenge this research. Chapter 5 examines the appeal and risks of using generational categories and provides alternative perspectives on the multiple influences on workforce development that can be applied in future research. Finally, Chapter 6 builds on the trends discussed in Chapter 2 to examine implications for workforce management. It highlights recruitment and retention challenges faced by the military, first responder, nursing, hospitality, and education sectors. In light of the committee's findings on the use of generational categories and the state of the generational literature, this chapter also summarizes legal constraints on workforce management and provides recommendations for employers on approaching management decisions and policy changes. This report also includes two appendixes: Appendix A summarizes the committee's approach to reviewing the literature on generational attitudes and behaviors in the workforce, while Appendix B provides biographical sketches of the committee members.

2

The Changing World of
Work and Workers

Workers carry out their jobs in the context of their workplace, and the workplace environment is affected by broader macrostructural forces. The context of work has changed along many dimensions over time, and one must understand these changes to examine properly any potential differences among workers. Context represents the situational opportunities and constraints at both the macro and mezzo levels that influence workers' interactions and behaviors; therefore, examining context is necessary to understand the degree to which changes in observed behavior and attitudes are due to such individual characteristics as generation. Understanding the context of work is also essential to identifying applications of research findings to workforce management (Johns, 2006).

Changes in work and workers have been the subject of frequent discussion among scholars, the popular press, management, and employees themselves (e.g., Hoffman, Shoss, and Wegman, 2020). This chapter summarizes the evidence on how work and workers have changed in both broad (macrostructural-level) and discrete (workplace-level) contexts (Johns, 2006). In so doing, it lays the groundwork for examining the intersection of these contexts with theories of and empirical research on generational differences in the workforce. Chapter 6 builds on the broad changes to work outlined here, and reviews specific challenges faced by employers in some job sectors and implications for workforce management.

THE BROAD CONTEXT OF WORK

Macrostructural changes in political, social, and economic institutions and structures, as well as in the broad context within which the military operates, are the basic drivers that have shaped the organization of work in the United States since the mid-1970s.[1] These changes are interrelated and together have altered the nature of employment and work arrangements in the United States, increasing employment insecurity, reducing workers' attachment to their employing organizations, and causing employers to recruit and seek to retain workers in new ways. (For overviews of these changes, see Cappelli, 1999; Cappelli et al., 1997; Hacker, 2006; Kalleberg, 2011; Osterman, 1999; and Osterman et al., 2002). These macrostructural changes have occurred in all developed countries, although their timing has somewhat varied (e.g., occurring later in Europe and the industrial countries of Asia than in the United States) (see Kalleberg, 2018). These macrostructural changes have affected both the private and public sectors.

Scholars in numerous social science disciplines, including sociology, economics, political science, anthropology, and history, have contributed to understanding the main macrostructural trends behind changes in the terms and nature of work and the characteristics of workers. Some of the significant macrostructural forces that have influenced changes in the economy, employer–employee relations, and the labor force are summarized below.

Changes in the Economy

Globalization

The global interconnectedness of production and finance, which are closely linked with international trade, has risen markedly since the end of World War II (Haskel et al., 2012). This period has seen substantial economic growth in many countries, particularly those occupying the middle of the spectrum of economies, including the "BRICS" nations of Brazil, Russia, India, China, and South Africa, which collectively accounted for nearly one-third of the world's gross domestic product (GDP) in 2015 (New Development Bank, 2017). As discussed below, the manufacturing sector has seen particularly large decreases in employment in the United States. A similar pattern of declines in manufacturing employment is found among other countries with advanced economies, including Canada, France, Germany, Italy, Japan, and the United Kingdom, while manufacturing

---

[1] Many researchers believe the drivers for the organization of work were different before the mid-1970s. They were still considered political, social, and economic in nature, but the characteristics of these structures were very different from those after the mid-1970s.

employment has increased in China and other rapidly developing middle-income countries.

Since 1980, rates of economic growth in Asia have surpassed those of Europe and North America;[2] this growth, combined with advances in communication technologies, has led to greater economic, political, and social interconnectedness among these three regions. A feature of this interconnectedness has been the offshoring of work from more developed countries to developing countries, where wages are lower and labor protections are weaker.

Technological Innovation

Advances in information and communication technologies have facilitated globalization of production and sped up product cycles on the one hand and introduced challenges of data management and security and worker education and training on the other (National Academies of Sciences, Engineering, and Medicine [NASEM], 2017a, 2017b). These advances are shifting the image of work in many sectors as a result of the automation of job tasks that can be routinized, the augmentation of workers' abilities to perform other tasks, and the creation of new kinds of jobs. Indeed, the pace of automation has raised renewed fears of a "jobless future" in which robots and computers will take over the jobs of vast numbers of workers (Autor, 2015). While researchers differ in their predictions about the extent to which this is likely to occur, it is important to keep in mind that extreme versions of this view are inconsistent with the evidence. The impact of information and communication technologies on employment will be determined not only by technical capabilities, but also by the choices made by policy makers, organizations, and workers in response to the economic, political, and social landscapes (NASEM, 2017b).

Continued Expansion of the Service Sector

The composition of employment in the U.S. economy has been changing over time. Historically, jobs have been classified by three major sectors defined by the nature of the work: extracting raw materials (e.g., agriculture, mining, and fishing); manufacturing products; and providing services. In 1970, 3.12 million Americans worked in the farming sector; by 2014 this number had declined to 1.95 million (Roser, 2020a). Employment in manufacturing also declined, from around 18 million in 1970 to 12 million in 2014 (U.S. Bureau of Labor Statistics [BLS], 2020a). At the same time, employment in the service sector expanded. See Figure 2-1. The U.S. Bureau

---

[2] See GDP Ranking at https://datacatalog.worldbank.org/dataset/gdp-ranking.

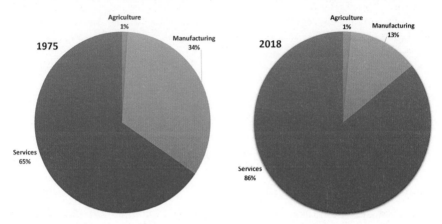

FIGURE 2-1 Employment by major industry sector 1975 and 2018.
SOURCE: Created by the committee with data from U.S. Bureau of Labor Statistics, QECW SIC-Based Data Files, CSVs, by Area, Annual Averages, 1975, https://www.bls.gov/cew/downloadable-data-files.htm and Employment Projections: Employment by Major Industry Sector, 2018, https://www.bls.gov/emp/tables/employment-by-major-industry-sector.htm.

of Labor Statistics projects continuing declines in manufacturing employment and further growth in service employment through 2026.[3] The service sector covers a range of employment in professional, personal, and public service, including but not limited to government, transportation, education, health care, and hospitality. The highest rate of growth is anticipated to be in health care and educational services. Notable growth in jobs in high-technology fields has also occurred since 2010 (The Computing Technology Industry Association, 2019).

In March 2020, in response to the COVID-19 pandemic, all sectors were immediately impacted, but certain industries were shut down (e.g., restaurants, hospitality, personal services), some had to ramp up quickly (e.g., health care and government), and many others had to adjust the way they do business. It has been estimated that 20 percent of all workers were employed in the industries that had to shut down temporarily (Dey and Lowenstein, 2020). The longer-term impacts of these transitions will be observable only over time.

Budget and Trade Deficits

The United States has had a budget deficit since the early 2000s, and the national debt has reached record levels (Desilver, 2019). The aging of

---

[3] See the U.S. Bureau of Labor Statistics' (BLS, 2019a) employment projections by major industry sector for 2008–2028 at https://www.bls.gov/emp/tables/employment-by-major-industry-sector.htm.

the U.S. population (see below), which will necessitate higher spending for social programs for the aged and contribute to a reduction in the proportion of the population that is active in the labor force, is anticipated to drive further increases in the U.S. budget deficit and national debt (Elmendorf and Sheiner, 2017). The U.S. budget deficit reflects a low rate of national saving, which, together with increasingly global financial markets, contributes to rising trade deficits (Cooper, 2008). These factors combine to create a greater risk of uncertain and poor economic conditions in the future.

Regulatory Environment

In the United States, a number of public policies instituted since the mid-1970s have emphasized market mechanisms and sought to provide key industries with greater flexibility through deregulation (beginning with airlines in 1978 and trucking and railroads in 1980), reduced enforcement of labor laws and standards, and overall replacement of government intervention in the economy with an enhanced role for markets (Harvey, 2005). These political decisions and the macrostructural economic changes described above were supported by ideological shifts in the U.S. culture toward greater individualism and personal responsibility for work (Bernstein, 2006), which represented a movement away from the idea that the government should provide economic security, as exemplified by the New Deal. Bernstein (2006) describes this vividly as replacing the idea that "we're all in this together" with the idea that "you're on your own." This shift toward greater individualism served as the normative basis for the massive deregulation of labor markets that occurred under the administration of President Reagan (Kalleberg, 2011).

Changes in Employer–Employee Relations

Shift in Corporate Governance

The 1980s saw a change in the conception of the firm from an entity that is committed to particular product markets and the production of goods and services (managerial capitalism) to one that is a bundle of assets to be bought and sold (finance capitalism). This "financialization" of the economy, under which capital markets play an increasingly important role in corporate decision making, was associated with a shift from the stakeholder model of corporate governance (which emphasized the welfare of managers, employees, customers, suppliers, and the community) to a shareholder model that gave priority to the interests of shareholders (Krippner, 2005). This shift put pressure on managers to increase profit margins and returns to shareholders and led to downsizing and outsourcing by even highly profitable firms seeking even higher profits.

Decline of Unions and Worker Power

The proportion of American workers who are union members has declined steadily since the 1950s (Hogler, 2020). This decline has been concentrated in the private sector, although recent antiunion legislation targeting public-sector unions enacted by governments in Wisconsin, Indiana, and Michigan has led to reductions in this sector as well. In 2018, 10.5 percent of the U.S. labor force comprised union members, down from 20.1 percent in 1983, with the union membership rate of workers in the public sector (33.9%) continuing to be much higher than that of workers in the private sector (6.4%) (BLS, 2020b). The continued decline of unions has helped shift the post-World War II balance of power from workers to employers. In turn, the weakening of worker power has facilitated the expansion of the macrostructural forces discussed above, such as globalization and changes in corporate governance. Moreover, the decline of worker power has helped accelerate the post-World War II reduction in institutional worker protections, which provided job security and contributed to the expansion of the middle class (see Greenhouse, 2019; Rosenfeld, 2014).

Changes in the Labor Force

Changes in the U.S. labor force since the 1970s have played an important supporting role in the macrostructural transformations discussed above. Among the key changes are the rise in the education levels and skills of the labor force, the growth in women's participation in the labor force, and evolving population characteristics (i.e., aging and increasing racial and ethnic diversity) (Fischer and Hout, 2006; Fry and Parker, 2018). In addition to general population trends that have affected the workforce broadly, certain lifestyle trends have implications for the workforce in some employment sectors (see Box 2-1).

Rising Educational Attainment

As noted, the United States has seen notable growth in educational attainment, which has occurred in conjunction with an increase in educational and economic inequality. Between 2000 and 2017, the percentage of people aged 25–29 with an associate's or higher degree increased from 38 to 46 percent, and the percentage with at least a bachelor's degree rose from 29 to 36 percent (National Center for Education Statistics [NCES], 2019). While the increase in educational attainment can fuel growth in

---

BOX 2-1
Lifestyle Trends

Two lifestyle trends—increasing urbanization and rising obesity rates—have implications for recruiting and managing workers for some job sectors.

Increasing Urbanization: In 1970, 74 percent of the U.S. population lived in urban areas; this proportion had increased to 81 percent by 2014 (Ritchie and Roser, 2020). Increasing urbanization has been a global trend over the past century. Since 2000, the majority of rural counties in the United States have seen more people move out than move in and a growing share of older adults in their populations (Parker et al., 2018). By contrast, urban communities have more diverse and younger populations. Therefore, workers in different areas have had varying exposure to different mixes of age groups and cultural experiences. Of note, there is preliminary evidence of recent shifts away from urban areas, which may offset earlier trends.

Rising Obesity Rates: The Centers for Disease Control and Prevention reports that in 2015–2016, 36 percent of people aged 20–39 were obese, as were 43 percent of those aged 40–59. The current rate of obesity among young adults is strikingly higher relative to previous generations and appears to be rising (Harris, 2010; Lee et al., 2010, 2011). The military is one employer that has been impacted by the rise in obesity, which has reduced the number of youth who are eligible to serve and raised concerns about future capabilities to protect the nation (Maxey, Bishop-Josef, and Goodman, 2018).

---

high-skill jobs, research indicates that many young people are "overeducated" for their jobs (Clark, Joubert, and Maurel, 2014). More than one-third of adults aged 18–29 have student debt; nationally, student debt totals $1.5 trillion (Cillufo, 2019). The percentage of postsecondary graduates taking loans to finance their education has risen since 2000 and was greater than 60 percent in 2016 (NCES, 2018).

Figure 2-2 shows the growth in the supply of more educated workers from 1963 to 2017. The share of the labor force consisting of those with college plus postcollege education increased from 12 to 39 percent, and the share of those with some college education from 13 to 28 percent; at the same time, the share of those with a high school degree or less declined from 75 to 33 percent. During this period, there was also a rise in the number of children growing up with a parent holding at least a bachelor's degree and the number of young adults pursuing a college degree after high school.

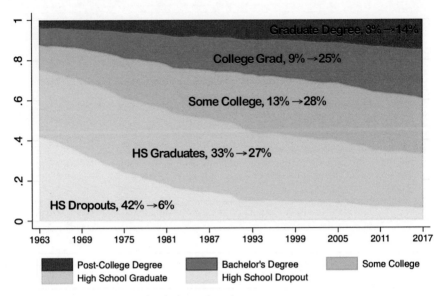

FIGURE 2-2 The rising supply of educated workers.
SOURCE: Autor, 2019b. Reprinted with permission.

Growth in the Employment of Women

The growth in the proportion of women choosing to work is a continuation of a longer-term trend. In 1950, as many as 34 percent of women were employed, compared with 86 percent of men (Toosi, 2002). The employment rate of women peaked at 60 percent in 1999 and was 57 percent in 2016. In contrast, by 1999 the employment rate for men had dropped to 75 percent, and in 2016 it was 69 percent (Hipple, 2016). As the employment rates of women have risen, so, too, has the percentage of dual-earner families and mothers as either sole or primary sources of family incomes. In 1970, 31 percent of households with children under age 18 had both parents working full time; as of 2015, this proportion had risen to 46 percent (Pew Research Center [PRC], 2015). In 1960, 11 percent of households with children were supported by mothers, either as single parents (sole providers) or primary providers; as of 2011, this proportion had grown to 40 percent (PRC, 2013).

The growth of two-career couples is one of many factors associated with rising urbanization (see Box 2-1), particularly among those with college degrees (Costa and Kahn, 1999). The rising rate of women's employment has not only transformed family structures but also affected the way people think about their jobs and consider what they need and want from paid employment, such as greater flexibility and control over work schedules

(Golden, Henly, and Lambert, 2013). This change has in turn created greater interdependencies between work and family; thus one cannot understand the consequences of changes in work environments without also taking into account family structures, and vice versa (Cherlin, 2014).

Population Aging

The U.S. Census Bureau projects that the number of Americans aged 65 and older will nearly double from 52 million in 2018 to 95 million by 2060, rising from 16 to 23 percent of the population. Population aging is a worldwide trend; between 2015 and 2050, the proportion of the world's population over age 60 will nearly double, from 12 to 22 percent (World Health Organization, 2018). Within the United States and worldwide, it is anticipated that rapid growth in the number of older people in the population will lead to greater demand for services and government-funded programs to meet their health and social care needs (Sullivan, 2016).

The aging of the population is also reflected in the increasing share of U.S. jobs that are held by people in older age groups. Retirement ages are rising: in 2002, the average retirement age was 59 years, while in 2014 it was 62 years (Riffkin, 2014). Continued employment of older people has contributed to the overall growth of the U.S. labor force and is one component of increased workforce diversity (discussed below). Figure 2-3 illustrates the trend away from a workforce dominated by young and middle-age workers toward one in which a growing proportion comprises older workers, aged 55 and up.

Rising Racial and Ethnic Diversity in the Labor Force

In addition to the age diversity noted above, the labor force has by many accounts become more diverse with respect to race and ethnicity (Fry and Parker, 2018). In 2018, according to the U.S. Bureau of Labor Statistics' annual report (2019b), "whites made up the majority of the labor force (78 percent).[4] Blacks and Asians constituted an additional 13 and 6 percent, respectively. American Indians and Alaska Natives made up 1 percent of the labor force, while Native Hawaiians and Other Pacific Islanders [constituted] less than 1 percent. People of two or more races made up about 2 percent of the labor force.... Seventeen percent of the labor force were people of Hispanic or Latino ethnicity, who may be of any race" (BLS, 2019b). This diversity of the labor force mirrors the increasing racial/ethnic diversity of the U.S. population generally (see Figure 2-4).

---

[4] Non-Hispanic whites made up 63 percent of the labor force in 2018.

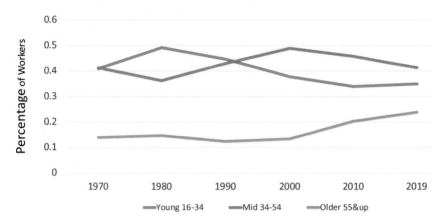

FIGURE 2-3 Proportion of employed workers by age in the U.S. labor force, 1970–2019.
SOURCE: Created by the committee with data from U.S. Bureau of Labor Statistics, Labor Force Statistics from the Current Population Survey (fourth-quarter figures, not seasonally adjusted).

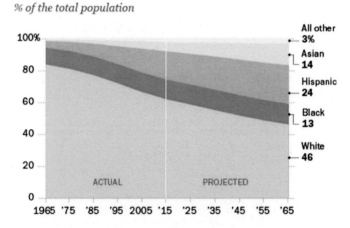

FIGURE 2-4 Racial and ethnic change in the American population.
NOTE: Whites, Blacks, and Asians include only single-race non-Hispanics; Asians include Pacific Islanders. Hispanics can be of any race.
SOURCE: Pew Research Center, 2015.

THE DISCRETE CONTEXT OF WORK

The macrostructural changes reviewed above have led to changes in the discrete context of work—the social and technical environments in which work is done, which include such aspects of a worker's immediate work environment as the types of occupations in one's workplace, financial incentives, the types of tasks performed, and the nature of business interactions.

The Increasing Complexity of Work

Social scientists have devoted considerable attention to documenting how the jobs and occupations that make up the economy have changed over time, and thus how the level of complexity associated with work has changed. This increased complexity reflects rapid technological advances, the emergence of the knowledge economy, and the expansion of the roles of employees—especially those in professional jobs—to meet competitive demands (Acemoglu and Autor, 2011; Fried, Levy, and Laurence, 2008; Morgeson and Campion, 2003; Wegman et al., 2018). According to a report of the National Research Council (1999), the cognitive demands associated with modern work, and knowledge-based jobs in particular, are much greater relative to previous decades. In addition, there is evidence that these changes have occurred both across and within occupations. In other words, although part of the observed increase in the complexity of work is associated with the increased proportion of high-skill occupations, evidence also indicates that the tasks of those high-skill jobs are more complex. Such competencies as problem solving and communicating continue to be important in many jobs, and the ability to critically evaluate and transfer knowledge is vital (Lyons et al., 2020). Further, the tasks workers must perform increasingly emphasize creativity, adaptability, and interpersonal skills over routine information processing and manual tasks (Wegman et al., 2018).

At the same time that the demand for high-skill labor has increased, advances in technology, among other macrostructural changes discussed above, have replaced many middle-skill jobs, resulting in an increase in the proportion of low-skill jobs in the economy.[5] The growth of the service sector has yielded continued demand for personal service jobs that are often low-paying. For example, there has been a surge in demand for personal care assistants to care for the aging population, which some have labeled "the care economy."

The rising rates of professional and technical employment with parallel falling rates of middle-skill jobs and rising rates of personal services is a phenomenon known as employment polarization. Changes in technology

---

[5] Scopelliti (2014) [p. 1.] notes that Cheremukhin (2014) "characterizes middle-skill jobs as routine jobs that are cognitive or manual in nature and require one to follow precise procedures; examples of middle-skill jobs with declining employment include cashiers and telemarketers (cognitive) and mail carriers and cooks (manual). He characterizes high-skill jobs as nonroutine and cognitive, requiring problem-solving skills—for example, analysts and engineers—and low-skill jobs, such as food service workers, as nonroutine and manual."

and automation and the decrease in institutional protections for middle-skill jobs have resulted in the polarization of jobs relative to skill requirements, especially in the 1990s. This represents a shift from the more monotonic employment growth of the 1980s, whereby occupational growth was slowest among lower-skill jobs and greater among high-skill jobs (see, e.g., Autor, 2019a, 2019b; Dwyer and Wright, 2019; Howell and Kalleberg, 2019). There is considerable debate among scholars as to when the trend toward employment polarization first began and whether it continues (Mishel, Schmitt, and Shierholz, 2013). Still, regardless of whether these trends are ongoing, the influences of the previous polarization of employment are still being felt by organizations and workers.

Rising Income Inequality

The macrostructural forces described above—especially greater globalization, rapid technological change, the financialization of the economy, and the decline of unions and institutional labor law and other protections—have led to the highest levels of income inequality seen in the United States since the early 20th century. The United States has seen declining average earnings across all education levels since 2000 even as the incomes of the highest-earning workers have risen markedly (Haskel et al., 2012). This income inequality has been coupled with growth in wealth inequality, including racial gaps (Keister et al., 2015). Residential segregation also has been increasing, and data suggest that the opportunities for Americans to move to higher-income categories have declined over the past several decades (Stanford Center on Poverty and Inequality, 2011).

Projections of future job growth portend ongoing inequality. According to the U.S. Bureau of Labor Statistics, 7 of the 10 occupations that are expected to add the greatest number of jobs between 2016 and 2026 pay less than the national median annual wage of $51,960.[6] In fact, four of these fast-growing occupations—personal care aides, food preparation workers, home health aides, and food servers—have median annual earnings below half of the median for all jobs. Although employment polarization has been blamed for growing income inequality, Hunt and Nunn (2019) recently demonstrated that employment polarization explains little of the growth in individual wage inequality.

Interdependence

As a result of the increase in jobs with greater cognitive complexity and the growth of the service sector, work has become increasingly

---

[6] See new-job projections at https://www.bls.gov/ooh/most-new-jobs.htm.

interpersonal and interdependent. Specifically, the flattening of once hierarchical organizational structures and the proliferation of teams seen in recent decades necessitate exerting lateral rather than downward influence to respond to volatile environmental demands more readily and achieve organizational objectives (Salas, Stagl, and Burke, 2004). Demand for interpersonal skills is evident not only in more complex jobs but also in low-skill jobs. Specifically, the growth of the care and service economies is associated with an increase in interpersonal interactions with customers and colleagues. Together, these trends point to the growing importance of interpersonal skills in the modern work world. In one of the few studies to examine this issue, Wegman and colleagues (2018) found that modern jobs require cooperation to a greater extent relative to jobs in the past.

Organizational Structure and Autonomy

The 1980s was an era of downsizing in which, when faced with increasing economic pressure from global competition and a shift to shareholder-focused business, organizations eliminated layers of midlevel management. To compete in the global marketplace, organizations also have restructured to decentralize decision making, thereby facilitating more agile responses to a turbulent business environment.

Autonomy refers to "the degree to which the job provides substantial freedom, independence, and discretion to the employee in scheduling the work and determining the procedures to be used in carrying it out" (Hackman and Oldham, 1975, p. 162). Global competition and the resulting flattening of organizational hierarchies have led to more diffuse decision-making authority, with lower-level employees being granted a greater span of control and more responsibility than in previous decades (Cappelli, 1999). This increased autonomy in decision making has resulted in demand for employees who can work independently and without supervision.

Finally, although the working hours of the average employee have been relatively constant since the 1970s, a greater proportion of the adult population is working (Rones, Ig, and Gardner, 1997), in part because of the increase in dual-earner families (U.S. Census Bureau, 2010). At the same time, aided by advances in information technology, there often is less need for workers to be in a central office location (Tam, Korczynski, and Frenkel, 2002). Autonomy in the form of flexible work schedules is meeting a need of employees trying to manage work and family roles and has become a more prevalent feature of the modern work environment (Wegman et al., 2018).

Organizational Fissuring

Organizations have increasingly been turning to contracting and subcontracting, resulting in complex relationships between workers and their employers and contracting organizations. For example, many workers are employed not by the organization at which they work but by other, contract organizations, a phenomenon sometimes referred to as "fissuring" (see Weil, 2014). As a result of fissuring, large corporations in particular have shed their traditional role as direct employers of the workers who produce their products and services, instead outsourcing these activities to smaller, contract companies. An important consequence is that contract company employees often receive less support and fewer benefits relative to workers in comparable jobs that are not outsourced because contract companies, in general, have fewer resources than larger companies. Further, contract companies often release larger companies from having to follow internal pay equity norms[7] (Howell and Kalleberg, 2019). As Appelbaum and Batt (2017, p. 77) summarize the literature on the topic: "Most empirical research in both the USA and Europe suggests that the rise of the networked firm and outsourcing of production has led to a deterioration in the jobs and pay of workers and to a growth in wage inequality."

Nonstandard Work

The "standard" employment relationship that was normative during most of the post-World War II period—in which employees worked for employers on a full-time, "permanent" basis at the employer's place of business and received regular pay and benefits—has been replaced as the employment norm in many cases by "nonstandard" work arrangements, such as temporary work, contract work, and independent contracting. Nonstandard work arrangements tend to be relatively uncertain and insecure and to lack many statutory and social worker protections (see Cappelli, 1999; Cappelli and Keller, 2013a, 2013b; Kalleberg, 2000). In such countries as the United States, many benefits, such as health insurance, are delivered via employers, and are often unavailable for nonstandard work arrangements.

While the evidence suggests that the percentage of workers in nonstandard work arrangements has increased only slightly since 1995 and still constitutes a minority of the labor force (see Howell and Kalleberg, 2019), the available information on nonstandard work is inadequate and likely underestimates this phenomenon, especially for workers who work as independent contractors to supplement their main jobs. Nonetheless, the

---

[7] When a client company (the "larger company") hires a contract company, agreements often allow that the client does not have to pay the contract company workers the same as it pays its regular workers for similar types of work.

qualities of nonstandard jobs are generally judged to be inferior to those of standard jobs, and evidence indicates that employment insecurity, low wages, and the shifting of risks from employers to workers increasingly characterize even standard employment (see Howell and Kalleberg, 2019).

Increasing Uncertainty

Because of the economic, political, and social changes discussed above, the world of work that was once more predictable and stable is now volatile, uncertain, complex, and ambiguous, or VUCA. The concept of VUCA was introduced by the U.S. Army War College in 1987 to describe dynamic threat environments, complete knowledge of which cannot be attained in the limited timeframe for decision making (Gerras, 2010; Jacobs, 2002). The VUCA concept has been adopted for business leadership (e.g., Bennett and Lemoine, 2014) and other purposes (e.g., sports management [Hogan, Santomier, and Myers, 2016]) as a way to frame the dispositions and skills required in a complex world.

Volatility is the rapid rate of change of the environment. Volatility in the Information Age means that even the most current data may not provide an adequate context for decision making. *Uncertainty* denotes the inability to know everything about a situation and the difficulty of predicting the nature and effect of change (the nexus of uncertainty and volatility). *Complexity* refers to the difficulty of understanding the interactions of multiple parts or factors and of predicting the primary and subsequent effects of changing one or more factors in a highly interdependent system or even system of systems. *Ambiguity* refers to the difficulty of interpreting meaning when context is blurred by such factors as cultural blindness, cognitive bias, or limited perspective.

DISCUSSION

This chapter has summarized some of the influential trends that have occurred in the United States in the broad and discrete contexts of work. Transformation in the broad or macrostructural context includes changes in the economy (e.g., shifts in the global economy, technological advances, and the dominance of service industries), changes in employer–employee relationships (e.g., declines in unions), and changes in the labor force (e.g., increased demographic diversity of employees with regard to gender, age, race/ethnicity, and education).

These macrostructural changes have been accompanied by changes in the discrete contexts of organizations and occupations. As a whole and over time, occupations have become more polarized, with a rise in high- and low-skill jobs and a decline in middle-skill jobs. As a result, there is demand for both highly skilled workers with advanced problem-solving skills and workers willing to take low-skill jobs. Further, for many occupations, critical job tasks have changed with the incorporation of technologies and the rise of nonstandard work, necessitating a review of the knowledge, skills, attributes, and other characteristics required of employees.

Taken together, the broad and discrete changes discussed in this chapter have led to greater uncertainty for both organizations and workers and greater insecurity for workers. Especially vulnerable are those workers who lack the human and social capital resources to achieve success in the labor market.

Many of the changes documented in this chapter have likely created new views of work, particularly for those just entering the workforce. These perceived differences between new hires and tenured workers may represent general adaptations to the changes in the broad and discrete contexts of work outlined in this chapter. As noted at the start, it is important to consider context when investigating differences among workers. The characteristics associated with a particular age group or generation of workers may reflect broad changes over time (i.e., period effects) in the nature of work rather than generational differences.

The evolution and use of generational theories are discussed in the next chapter, followed by a review in Chapter 4 of the state of the research that has sought to identify generational differences among workers. This review considers whether this research has sufficiently separated generation effects from period or age effects. Later in the report, Chapter 6 revisits the implications for workforce management of some of the trends discussed here.

> **Conclusion 2-1:** Understanding potential and meaningful differences among workers requires consideration of the broader context of work. Research across disciplines has identified shifting economic, military, political, and societal trends that have led to workplace and workforce changes. For example, technological advances, globalization, and other factors are altering the nature of work. In addition, the diversity of the labor force has increased in terms of age, gender, race, and ethnicity. To remain competitive, organizations are facing the need to align their workforce management policies and practices with the changing world of work.

3

Origin and Use of Generational Theories

This chapter reviews the history and use of generational theories, as well as the creation of such generational labels as "baby boomers" and "millennials" that are used to describe people born in a certain time period or of a certain age. While the notion of generations has a long history of scholarly consideration, attention to generational differences among individuals has become increasingly prevalent over the past 20 years. An implicit assumption of generational thinking is that people who were born around the same time have similar values and attributes that differ from those of people born at a different time. This chapter provides background on how the concept of generation is used and has evolved in both the scientific and popular literature. The next chapter reviews the existing scientific literature related to generational differences in the workforce.

EARLY SOCIOLOGICAL THEORIES OF GENERATIONS

In the mid-1800s, Auguste Comte described social progress as the product of generational change (*Cours de Philosophie Positive*, 1830–1840). He posited that just as individuals mature and change throughout their lives, societies progress through stages, which he termed "theological," "metaphysical," and "positive." According to Comte, progress through these stages is driven by generational turnover, with each successive generation bringing new and innovative ideas and practices to replace those of older generations.

The modern scientific usage and understanding of the term "generations" can be traced back to sociologist Karl Mannheim's *The Problem of Generations* (1952). Mannheim theorized that generations provide a basis for understanding social movements—how social change is possible while cultural traditions and identity are preserved. He identified five processes through which generations facilitate social change: (1) new participants in the cultural process emerge, (2) former participants in the cultural process disappear, (3) members of any generation can participate only for a limited time, (4) cultural heritage is transmitted from generation to generation, and (5) generational transitions are continuous. Subsequent sociological theories similarly highlighted the importance of generations in facilitating social change. Ryder (1965), for instance, described the succession of birth cohorts (a construct similar to Mannheim's formulation of generations) as a process of lending flexibility and providing new perspectives to address social problems.

According to Mannheim (1952), generations are formed through two important elements: a common location in historical time, such that there are shared events and experiences, and an awareness of that historical location. Mannheim clarified that a generation is not a "concrete group" of people who share physical and social proximity and are aware of the existence of the other members. Thus, a generation is similar not to a club, in which one could identify who is in and out, but to a person's social class. Notably, Mannheim stressed that birth year alone was insufficient to place a person in a specific generation; rather, the person needed to experience and participate in the defining events of the generation. He also noted that the same historical events will not affect people from different cultural backgrounds and social classes in the same way. To use a modern example, the destruction of the World Trade Center Towers in New York City on September 11, 2001, might be expected to have significantly affected the thinking and attitudes of Americans who were alive to witness these events. But such an event might not be a defining moment for people living in countries in which terrorism is more frequent than is the case in the United States, and it would be expected to affect the thinking and attitudes of New Yorkers differently from those in other parts of the country. Even within the United States, people of different education levels, wealth, and culture may have experienced or interpreted these events differently.

Both Mannheim (1952) and Ryder (1965) rejected the idea that generations emerge at regularly spaced intervals, noting that the rhythm of generations depends on the timing of historical, social, and cultural events that affect people's experiences. Ryder further noted that historical events occurring during young adulthood are particularly influential, as young people are "old enough to participate directly in the movements impelled by change, but not old enough to have become committed to an occupation, a residence, a family of procreation or a way of life" (Ryder, 1965, p. 848). Later generational theories in sociology highlighted the importance of not only historical events that happen during especially salient developmental stages, but also significant culturally bound life stages (e.g., education, marriage, building family, working years) that influence goals and values (Riley, 1973, 1987). These sociological theories of generations did not focus on understanding individual behavior, but on an aggregate concept of generations as facilitating social change (Rudolph and Zacher, 2017).

Another major figure in the sociological tradition is Glen Elder. Building on his large-scale longitudinal studies of child and adult development produced during the early to mid-1900s (Elder, 1974, 1985), Elder formulated a life course perspective (Elder, 1998; also see Elder, Kirkpatrick-Johnson, and Crosnoe, 2003) positing the process through which social and historical contexts, particularly during childhood and adolescence, affect the trajectory of an individual's development through the life span. Specifically, he argued that "historical events and individual experience are connected through the family and the 'linked' fates of its members" (Elder, 1998, p. 3). That is, an individual's childhood and adolescent experiences are critically important in setting the stage for the subsequent developmental adult trajectory. In contrast to sociological traditions emphasizing the impact of events on social change, Elder focused on the mechanisms and consequences of social and historical context with respect to an individual's values and transition into adult roles, most notably those related to work.

Elder argued that an individual's family resources, values, and strategies for adapting to the broader external context exert a stronger effect on that individual than the historical context per se. Thus, within a generation or cohort defined by historical period, one could expect great heterogeneity within that population segment as a function of both more proximal familial and social interdependencies. In a similar vein, MacLean and Elder's (2007) review of the literature shows that the effects of different historical periods on military service are moderated by person-related attributes (e.g., family and friend resources).

Elder's life course perspective extended the sociological approach to generations in two ways. First, by focusing on the individual and lifespan development, his ideas helped shift attention from impacts on social change to impacts on individual behavior and adult development. Second, along with the work of Riley (1987) and others, Elder's concept of "linked lives" emphasized a possible process or mechanisms by which unique events that often characterize a generation come to affect an individual's values and behavior (Elder, Kirkpatrick-Johnson, and Crosnoe, 2003). However, Elder consistently noted that the variability associated with different "linked lives" in turn yields nontrivial variability in how individuals who live through a similar time period develop different values, interests, and occupational trajectories.

INFLUENTIAL POPULAR THEORY OF GENERATIONS

Like the researchers of the sociological theories described above, Strauss and Howe (1991) focus on generations in the aggregate in their popular book *Generations: The History of America's Future 1584 to 2069*. Their approach, however, departs from scientific theories in two important ways. First, they delineate a specific span of time—about 20 years—associated with the emergence of a generation. Second, they posit that four generational personalities (idealist, reactive, civic, and adaptive) emerge every 20 years or so in a cyclical pattern that repeats roughly every 80 years, driven by a generational reaction to the prior generation. According to Strauss and Howe, for example, idealists are an indulged and narcissistic generation of adults who raise a generation of underprotected and alienated reactives; who then raise team-oriented, overprotected but society-minded civics; who then raise an adaptive generation that comes of age in a time of crisis with an ethos of personal sacrifice. Although this pattern supposedly repeats every 80 years or so, the authors allow for significant historical events, such as the Civil War, to disrupt the cycle.

Although their work is thought-provoking, Strauss and Howe (1991) present essentially no empirical evidence for their theoretical perspective. Rather, they highlight individual case studies in making their claims regarding prototypical representatives of each generation's personality type throughout history. Although these case studies are compelling, they were selected specifically to provide examples of prototypical members of a generation (i.e., selection bias). One might also easily provide counter examples of people within a cohort who do not exhibit the prototypical traits associated with a generation or who exhibit traits belonging to a different generation (e.g., people who should be adaptive given their birth year but who exhibit the traits of an idealist). Nonetheless, the work of Strauss and Howe has been highly influential with respect to both their thinking about the timing of the emergence of generations (i.e., every 20 years or so) and their labels for generations, which have influenced popular ideas about generational differences (Brooks, 2000).

GENERATIONAL LABELS

In the modern era, generations are often described by labels and defined as a group born between specific years—for example, the "millennial" generation, born roughly in the 1980s and 1990s. Generations tend to be assigned these labels through a somewhat messy process led by journalists, magazine editors, advertising executives, and the general public (Raphelson, 2014). Usually, a variety of labels are used until one sticks in the common vernacular, because of either a seminal book or article, a historical event, or simply general consensus. As noted above, Strauss and Howe's (1991) work was highly influential with respect to how it described and labeled the various generations in America. The labels they used for each of the generations have become—with one exception—the common vernacular in discussions about generations. Although they did not create the terms, their labels of the "silent," "boomer," and "millennial" generations have stuck.

The term "silent generation" notably appeared in a 1951 article in *Time* magazine, but it is unclear when the term originated.[1] The labels "baby boomers" and "millennials" are linked to historical events, but it is also not entirely clear who created these terms or how they came to be the "official" names for those generations. "Baby boomer" was given to the generation of individuals born between mid-1946 and mid-1964, after World War II (Hogan, Perez, and Bell, 2008). The term denotes the baby boom in the United States following the war, when the birth rate rose significantly and then fell. The label was notably used in a 1963 newspaper article about the new wave of college applicants.[2] The term "millennials," referring to the turning of the millennium, appears to have first been coined by Strauss and Howe.

"Generation X" (called the "13ers" by Strauss and Howe) is a striking example of how generational labels are largely the product of popular culture. Photographer Robert Capa first used the title *Generation X* in the 1950s for a photo series of young people after World War II. In 1964, a collection of interviews with teenagers was published in a book titled *Generation X* (BBC News, 2014). The phrase was again used by musician Billy Idol in the early 1970s for his punk rock band. The label finally achieved its modern meaning after being popularized in a 1991 novel by Douglas Coupland, a Canadian author and artist. Interestingly, Coupland's choice of the title *Generation X* was meant to signify that this generation did not want to be defined (Raphelson, 2014).

Finally, the popular label "generation Z" is used by many writers for the youngest named generation. It recently appears to have won out over other contenders (Dimock, 2019), such as "postmillennials" (Fry and Parker, 2018), "iGen" (Twenge, 2018), and "homelanders" (Howe and Strauss, 2007).

---

[1] See the "People: The Younger Generation" piece in *Time* (November 5, 1951) at http://content.time.com/time/subscriber/article/0,33009,856950,00.html.

[2] See clip from Daily Press (Newport News, VA, January 28, 1963) at https://www.newspapers.com/clip/19690752/daily_press.

WIDESPREAD USE OF GENERATIONAL TERMINOLOGY

The topic of generations and generational differences is discussed in a wide variety of contexts, including the popular press, business and human resources advice, and research,[3] both in the United States and internationally.[4] In these contexts, authors and consultants make use of the generational labels and associated birth years as an easy way to categorize groups of people, primarily by age (see Table 3-1 for examples). That is, generational categories are used commonly as a heuristic to reference a group of people around a certain age.

The use of generational categories in discussions about workforce management has become particularly prominent in the past 20 years in the popular press and in businesses and human resources advice. This growth in the use of these categories with respect to workforce management suggests anecdotally that employers are taking a serious look at generational differences. While the committee could find no evidence of enacted employment policies and practices directly tied to generational issues, we did find opinion pieces, commissioned reports, and training aimed at addressing personnel concerns from a generational perspective (see Box 3-1 and the discussion below).

Popular Press

It was beyond the scope of the committee's charge to review comprehensively the coverage of generational issues in the popular press. In conducting this study, however, we could not help but notice the vast amount and array of advice on generational issues in the workforce that is available to the public. Here, we offer our observations after reading many of these articles.

---

[3] Government agencies that collect population-level data, often used by researchers and the public, sometimes report these data by age groups using generational categories. Notable agencies include both the Department of Labor (BLS, 2019c) and the Census Bureau (U.S. Census Bureau, 2015).

[4] Given the claim that generations are formed through shared experiences of events that occur during a developmentally significant period (i.e., late adolescence/early adulthood), it is curious that labels for generations that were generated in the United States have also been used to describe and explain behavior for people and cultures outside of the United States, who arguably do not share the same cultural experiences. For example, Pew Research draws the line between millennials and generation Z as 1996 based on the timing of a few events: the terrorist attacks of September 11, 2001, U.S. involvement in the wars in Iraq and Afghanistan, the 2008 election and economic recession, and the adoption of such technologies as the smartphone (Dimock, 2019). However, these events, which mark the lines between generations in the United States, would have been experienced differently or not at all in other countries around the world.

TABLE 3-1 Illustration of Different Labels for Generational Categories and Associated Birth Years from Various Sources

Relative Age of Worker in 2020	Fry (2018)	Howe and Strauss (2007)	Campbell, Twenge, and Campbell (2017) and Twenge Website*	Variations in Birth Years among Researchers (Costanza et al., 2012)
Under 25	Postmillennial or generation Z (1997 or later)	Homeland (2005–2025?)	iGen (1995–2012?)	
26–40	Millennial (1981–1996)	Millennial (1982–2005?)	Millennial (1980–1994)	Millennial (1976/1982–1999/2000 or later)
41–55	Generation X (1965–1980)	Generation X (1961–1981)	GenX (1965–1979)	Generation X (1961/1965–1975/1981)
56–74	Baby boomer (1946–1964)	Boom (1943–1960)	Baby boomer (1946–1964)	Baby boomer (1943/1946–1960/1969)
75 or older	Silent/greatest (1945 or earlier)	Silent/GI (1942 or earlier)		Silent (1945 or earlier)

* See FAQs *What are birth year cutoffs?* at http://www.jeantwenge.com/faqs.

Most articles in the popular press, as well as television news stories, that refer to generations report on the likes, dislikes, habits, and attributes of various generations. For example, a Google news search of the word "millennial" yields more than 50 million results, with articles and stories from the *Los Angeles Times*, *Forbes*, NBC News, *The Washington Post*, and many other sources. The topics of these articles and stories range from a generation's thoughts about religion and pets to their values and behaviors in the workplace. Much of the discussion of these topics in the popular press is descriptive in nature and reports demographic statistics, such as the percentage of millennials in the workforce.

Most of the research and data referenced, if any, in these pieces is cross-sectional, drawn from either published cross-sectional research or some informal survey conducted by the reporting outlet. Findings from these surveys can be misleading because the findings are often reported as generational characteristics but may just be reflecting age differences at the time, and because of the small convenience samples on which the findings are based are not representative of the generations the surveys seek to characterize. Issues of cross-sectional designs and representativeness are discussed further in Chapter 4.

BOX 3-1
Employer Example: Is the Military Concerned
about Generational Differences?

The military employs people of a wide range of ages, from new enlistees in their late teens to senior flag officers in their 60s. Because it recruits young adults almost exclusively, however, the attitudes and behaviors of young people are of keen interest to the military. This interest may translate into concerns about how best to recruit and manage the "new generation," as well as concerns about managing an intergenerational workforce.

Several public documents address generational concerns in the military. Wong (2000), for example, examined differences between generation X and baby boomer officers in the Army. That analysis led the author to recommend several courses of action the Army could take to appeal to younger officers, including giving officers more time for family; making "the Army a fun place to work and live"; and "encouraging advanced civil schooling, training with industry, or sabbaticals" (pp. 19–20). The author's evidence for generational differences is based on survey responses from two cohorts of officers: baby boomer captains in 1988 and generation X captains in 1998. As Wong himself acknowledges, this kind of comparison conflates general changes in workers over time with generational differences (as discussed further in Chapter 4).

In 2007, the Department of Defense, through its 10th Quadrennial Review of Military Compensation, commissioned the Center for Naval Analyses to "conduct background research on millennials…to explore the potential impact of targeted policies, especially compensation and retirement, on this cohort" (Stafford and Griffis, 2008, p. 1). Based on a review of the literature, employment practices, and other data sources, the authors conclude that people within a generation will vary with respect to a number of experiences and expectations, that young people will mature and change, and therefore that effective workforce policies will consider a broader set of characteristics beyond generation.

Despite this pushback on a generational perspective, generational concerns reemerge with each new generation of military recruits, as evidenced by a set of commentaries in military periodicals (Cunningham, 2014; Reid, 2018), a recent report on talent management (Army Science Board, 2015), a recent bibliography of resources on generational differences and age discrimination (DeBickes and Stiller, 2016), and a training course on managing millennials.*

---

* See the Marine Corps training on Managing Millennials at https://www.hqmc.marines.mil/hrom/Sponsored-Training/Course-207.

As the group of individuals known as the millennials have aged (now in their 20s and 30s) and become a dominant proportion of the global population, an entire industry on generational differences has developed in an effort to understand the expectations of this target group and capitalize on those expectations economically. In many cases, generational labels are presented as heuristics with which to better understand differences in work-related values and other attitudes of different age groups. Many of these articles use headlines to highlight large differences among generations, but the further one reads, the more generational differences are described as somewhat trivial. Moreover, many authors include caveats that highlight either heterogeneity within generations (e.g., not all millennials are always on their cellphones) or evidence showing similarities between generations (e.g., millennials and generation Xers use their cellphones equally). The committee observed trends in the popular press of "myth busting" some generational claims, reporting discord among individuals who feel they do not belong to or identify with common stereotypes of a given generation, as well as growing instability in the concept of easily generalizable groups based on either birth year or shared historical events (e.g., Casey, 2016; *Wall Street Journal*, 2017).

Business and Human Resources Advice

A plethora of discussion and advice concerning generations in the workplace is available in books, magazine and newspaper articles, blogs, and surveys and from a growing number of consultants who provide training and perspective on these issues. For example, Deloitte, a large international consulting firm, conducts an annual "Millennial Global Survey" to look at attitudes, perceptions, and characteristics of young people around the world.[5] This survey is administered to around 10,000 people from dozens of countries, all of whom were born between 1983 and 1994, and includes questions related to work, the economy, technology, and similar issues. Business-centered organizations and publications have numerous articles and courses on managing different generations. For example, there are hundreds of articles about generations in the workplace on the website of the Society for Human Resource Management (SHRM), and the American Management Association offers articles and several courses about managing

---

[5] The Deloitte Global Millennial Survey uses the same birth cohort (1983–1994) to examine "millennials" from 42 countries, including Nigeria, Australia, Malaysia, China, and South Africa. Deloitte is not alone in applying U.S.-based generational categories to the global population (despite the definitions linking generations to significant social events and lack of relevance to international populations). Numerous popular articles report on work-related characteristics of people around the world using such U.S. categories as generation X, millennial, and generation Z (Bresman and Rao, 2017; Miller and Lu, 2018).

generations. There are even a handful of business school courses centered around issues related to generations.

Much of this advice focuses on the challenges of managing workers of multiple ages in the workplace and often includes broad descriptions of each generation, with little reference to evidence supporting these descriptions. For example, an article on the American Management Association website describes the silent generation as loyal and dedicated, baby boomers as distrusting authority and having a sense of entitlement in the workplace, generation X'ers as independent and placing a lower priority on work, and millennials as resilient and team-centric (Jenkins, 2019). While the article does not explicitly acknowledge potential heterogeneity within generational groups, it goes on to say that employers should seek to "create a respectful, open and inclusive environment where workers of all ages and cultural backgrounds can share who they are without fear of being judged, 'fixed,' or changed." Other articles use headlines that appear to claim large differences among generations, but the articles themselves often state that workers of all ages generally want the same things out of work. For example, the main message of an SHRM article titled "Employers Say Accommodating Millennials Is a Business Imperative" is that workplaces in which flexibility, work–life balance, and wellness are emphasized are able to attract and retain workers of all ages (Wright, 2018). Taken as a whole, this advice often is self-contradicting, identifies similar values among workers (e.g., seeking respect on the job and personal growth), or boils down to the assertion that workers should be assessed individually and not by generational group.

Research

As the idea of generational differences has grown in popularity, so, too, has the number of studies in this area by think tanks, scientific organizations, and researchers. The Pew Research Center, a nonpartisan think tank, collects and analyzes data on such issues as work attitudes, use of technology, and economics; the Pew website lists hundreds of articles relating to age and generation, some of which look at data using age categories and some of which use generational categories.[6] The American Psychological Association (2017) conducts an annual survey on Work and Well-Being, which compares generational groups on a number of work outcomes, including work stress, job satisfaction, and plans to change jobs. The Society for Industrial and Organizational Psychology (SIOP) has published white papers on the millennial culture in the workplace (e.g., Graen and Grace,

---

[6] See resources from the Pew Research Center at https://www.pewresearch.org/topics/generations-and-age.

2015) and has featured such talks as "What Millennials Want from Work" at its annual conferences. The information provided by these scientific organizations, however, also tends to include perspectives that examine the quality of evidence behind generational stereotypes (see, e.g., the September 2015 special issue of *Industrial and Organizational Psychology: Perspectives on Science and Practice*, Volume 8, Issue 3).

Psychological research on generations typically attempts to link such individual outcomes as work-related values, attitudes, and behaviors to generation membership. Notably, this research tends to use existing generational categories (see Table 3-1) to define groups in its samples. The committee was tasked to review the body of literature on generational attitudes and behaviors in workforce management and employment practices, and we identified more than 500 research articles on the topic. Appendix A details the committee's literature review, while Chapter 4 presents the findings and conclusions that resulted from the review.

SUMMARY

Research has explored the concept of generations for decades as a way to understand social change. New approaches have taken ideas from the sociological literature and applied them to understanding attitudes and behaviors of individuals. The idea of categorizing people by their generation became popular, and generational terminology has now taken hold in the common vernacular. Numerous articles and discussions and a growing industry of consultants and management resources focus on generational differences and the management of generations in the workplace, and employers and managers are being urged to make decisions and develop policies based on generational differences. However, careful examination of the empirical support for generational differences is essential before significant, costly decisions are made. Findings from existing generational research on work-related outcomes are examined in the next chapter.

Conclusion 3-1: As popular use of generational terminology expanded, the concept of generations developed decades ago in sociology to understand social change has taken a new research trajectory in an effort to classify individual differences in values, attitudes, and behaviors, notably those in relation to work. This new trajectory has been fueled in the past 20 years by greater attention to changing workforce demographics and the potential utility of understanding generational differences with respect to work.

4

Review of the Generational Literature

The committee was tasked to "gather, review, and discuss the business management and the behavioral science literature on generational attitudes and behaviors in workforce management and employment practices." As discussed in Chapter 3, research on generations has a long history, but attention to generations and work-related attitudes and behaviors is fairly recent, with empirical studies notably on the rise in the past 20 years (Costanza et al., 2017). In its search for relevant scientific literature, the committee identified more than 500 articles that have been published since 1980 (see Appendix A for detail on the committee's search strategy and the literature identified for this review). This chapter summarizes our findings and conclusions about the state of this body of research, referred to collectively here as generational research or generational literature. In our review, we drew on findings from previous reviews of this literature and critiques of the dominant methodologies used in these studies, and we conducted a pilot review of a small subset of the articles to confirm our findings.

OVERALL STATE OF THE LITERATURE

Since the National Academies letter report (National Research Council [NRC], 2002) discussed in Chapter 1 was published nearly 20 years ago, the amount of empirical research on work-related generational claims has increased considerably. Several scholars have noted the paucity of empirical studies published before the late 1990s (Costanza et al., 2017; Parry and

Urwin, 2011), and the committee observed this as well: the majority of articles we identified were published after 1999 (see Appendix A).

As the idea of generational differences in the workforce has grown in popularity (see Chapter 3), new lines of inquiry, based primarily in the disciplines of psychology and business management, have adopted early sociological theories on generational shifts and social change (Mannheim, 1952; Riley, 1987; Ryder, 1965) as a framework for characterizing individual attitudes and behaviors. Seeking to verify and/or identify generational differences, empirical studies have been conducted to measure work-related attitudes and values, often with the assumption that attitudes and values directly influence behaviors in the workplace. Very few studies focus specifically on actual workplace behaviors[1] as this kind of data is difficult to collect. Instead, most research is based on self-reported responses to surveys. Moreover, research has varied greatly with respect to the questions (i.e., item responses) and length of surveys used to operationalize values or attitudinal variables. In general, many of the values measured can be categorized in terms of work ethic, work centrality and leisure, altruistic values, and extrinsic values (Twenge, 2010). With regard to attitudes, most empirical studies focus on job satisfaction, organizational commitment, and intent to leave (Costanza et al., 2012; Parry and Urwin, 2011). Other scholars have remarked on the paucity of studies examining generational differences as regards training, motivation, and leadership (Rudolph, Rauvola, and Zacher, 2018). There is, however, much research examining the relationship of age to work motivation (Inceoglu, Segers, and Bartram, 2012; Kooij et al., 2011), and job performance (Ng and Feldman, 2008, 2010b). Further discussion of the literature related to the aging workforce is included in Chapter 5.

Much of the generational work has been undertaken with the purpose of helping organizations better understand how to recruit, develop, retain, and/or motivate individuals from different generations (Williams, 2019). Some of this work has been undertaken to test the very notion of generational differences, often testing assumptions or findings of other studies by using more sophisticated methods or data sources (Costanza et

---

[1] Some studies have tried to better understand generational differences in regard to job turnover, but most of these use survey responses in relation to intents to leave job as opposed to actual behaviors of switching jobs. One study (Enam and Konduri, 2018) did examine changes in time engagement behaviors over time and found that those born between the years 1982–2000 were more likely than people born earlier to delay entry into the workforce and exhibit longer student status in higher education. This study did not look at the time engagement behaviors of the subsequent generation so it is not clear whether this finding is specific to "millennials" or part of a more general upward trend toward more education and later entry into work.

al., 2017; Parry and Urwin, 2017; Twenge, 2010). The latter studies have documented the conceptual and methodological limitations of the former:

- "Although in some areas (e.g., work centrality) time-lag and cross-sectional studies are fairly congruent, in other cases they disagree. Where they are discrepant, the most logical explanation is that the cross-sectional study is also tapping differences due to age or career stage.... The other possibility is that the time-lag studies are finding a time period effect, i.e., all generations have changed over time in the same way" (Twenge, 2010, p. 206).
- "... current [cross-sectional] approaches adopted for the investigation of generations across most studies are fundamentally flawed.... Any study, whether quantitative or qualitative, that only considers a group of individuals at one point in time is unable to distinguish between age, period, and [generation] effects.... The only way to achieve any insight into these three different effects is to investigate generational differences using longitudinal data.... [Unfortunately,] there are limited numbers of datasets that (either quantitatively or qualitatively) ask the same questions of individuals, of different ages, as part of a panel or repeated cross-section, over decades" (Parry and Urwin, 2017, p. 142 and 146).
- "... none of the three commonly used approaches [in the generational literature] fully and accurately partitions the variance to age, period, and cohort in a generations context.... The linear dependency created when defining generations as the intersection of age and period creates an unresolvable identification problem, making it very difficult to isolate the unique effect of any one of the factors" (Costanza et al., 2017, p. 161).

Given these conceptual and methodological issues, discussed further below, it is exceptionally difficult to draw firm conclusions about generational differences in work-related variables. The current state of evidence suggests that any observed differences among workers are more likely to reflect age differences at the time of measure or evolving social and work conditions as a result of historical events impacting all people (see Chapter 2) than true generational distinctions. Further, the few studies that have used datasets with measures of attitudes over time have found weak generation effects, indicating that the variability in work-related attitudes and behaviors within generational groups is likely to be larger than the variability among generations. That is, individuals from the same "generation" are just as likely to be different from one another as from individuals of different generations.

Conclusion 4-1: Many of the research findings that have been attributed to generational differences may actually reflect shifting characteristics of work more generally or variations among people as they age and gain experiences.

CONCEPTUAL ISSUES IN THE LITERATURE

Research on generations has varied across disciplines, and the use of the term "generation" has typically fallen into two different perspectives: one focused on descent and lineage-based linkages, prominent in such fields as anthropology, and the other on shared experiences of a group of people of similar age, more prominent in sociology (Burnett, 2011; Joshi, Decker, and Franz, 2011). Much of the generational literature on work-related variables discussed in this report draws on early theories in sociology (Mannheim, 1952; Riley, 1987; Ryder, 1965) regarding generations and social change (see Chapter 3). Instead of studying social change, however, the bulk of the work-related generational research adapts these sociological theories to the study of individual attitudes and values. Three elements of these early theories have been used as a basis for recent empirical studies: (1) significant events occur within a society that broadly affect a societal group of individuals; (2) these events impact a particular cohort—a birth cohort in their formative years of late adolescence/early adulthood; and (3) events that happen during people's formative years exert a continuous influence on their thoughts and behaviors as they age—hence the emergence of a generation (Parry and Urwin, 2017).

The scientific literature contains many variations of the definition of the concept of generations, based in part on the conceptualization of Mannheim.[2] These include

- "an identifiable group that shares birth years, age, location, and significant life events at critical developmental stages" (Kupperschmidt, 2000, p. 66);
- "a cultural field in which social agents participate to varying degrees dependent upon their structural location within society" (Gilleard, 2004, p. 117);
- "a group of individuals, who are roughly the same age, and who experience and are influenced by the same set of significant historical events during key developmental periods in their lives, typically late

---

[2] In the seminal work of sociologist Karl Mannheim, "The Problem of Generations" (1952), "generation" suggests a group of individuals of a similar age and a similar location who experience similar social, historical, and life events (Lyons and Kuron, 2014; Parry and Urwin, 2011).

childhood, adolescence, and early adulthood" (Costanza et al., 2012, p. 377); and

- "[a group] of individuals born during the same time period who experience a similar cultural context and, in turn, create the culture (Gentile, Campbell, and Twenge, 2013)" (Campbell et al., 2015, p. 324).

In conducting empirical research, such theoretical concepts as generations need to be operationalized so they can be linked to variables that can be measured and studied. The definitions above suggest that the concept of generation is a complex mix of age, location, and context; however, it rarely is operationalized as such. The concept has been difficult to operationalize, and in many studies, birth cohort is used as a proxy for generation (Brink, Zondag, and Crenshaw, 2015; Parry and Urwin, 2011). For researchers, it is straightforward to classify individuals by birthdates but much more difficult to know when those individuals of the same birth cohort were also exposed to the same the sets of experiences. Many resources, popular and academic, add to the confusion by using the concept of generations interchangeably with that of age groups. This is particularly noticeable in resources that draw solely on one-time cross-sectional studies (see the discussion below).

The approach in most studies reviewed by the committee is to take predefined cohorts based only on birth year as representing distinct generations (Parry and Urwin, 2011). The labels and a range of birth years for each generation are generally assumed (see Chapter 3); across studies, however, there is substantial variation as to the exact starting and ending year for each group (see Costanza et al., 2012; Rudolph, Rauvola, and Zacher, 2017). For example, the label "baby boomers" is often used to apply to people born between 1946 and 1964, although, depending on the study and its criteria for categorizing research subjects, this range of birth years can be shorter or longer. As a result, generational categories are inconsistent across studies. Moreover, this lack of consensus on birth years for different generations indicates that there has been no empirical justification for any birth-year boundaries.

Using just birth years to define cohorts assumes that the influence of proximal historical events and social, cultural, and economic phenomena on those cohorts' individual members has already been established—an assumption largely untested. Thus, the generational research tends to take as antecedent an undefined set of shared experiences assumed to have shaped the attitudes/values to be measured. While a few articles make reference to what those influences might be (e.g., growing up with new technologies or the lasting influence of significant events during adolescence, such as war, the moon landing, or the terrorist attacks of September 11, 2001), none of them defines or investigates the mechanisms through which a specific event or phenomenon directly shapes the variable of interest (e.g., job satisfaction, work centrality). Because this research has generally adapted existing theories and conceptualizations of generations without examining assumptions built into the definition of generational groups, its theoretical contribution has been somewhat limited.

METHODOLOGICAL ISSUES IN THE LITERATURE

In addition to the theoretical issues of defining generations and specifying the precise mechanisms that differentiate them, other methodological issues make generations a challenging topic to study: the data needed to address generational questions related to the workplace rigorously are often difficult to obtain, and appropriate statistical approaches for studying generations are complex. This section examines some of these methodological challenges of separating out age, period, and cohort effects, and then reviews analytic approaches currently used to draw generational inferences and the concerns they raise. It also looks at other methodological concerns involving measurement and sampling.

Challenges of Separating Age, Period, and Cohort Effects

Research aimed at identifying and determining the extent of generational influences in the workplace essentially tries to separate generation effects from age or period effects (see Box 4-1). These three concepts and the statistical and methodological challenges associated with isolating the influences of each are present in large bodies of work in the fields of demography, economics, epidemiology, political science, psychology, and sociology (see, e.g., Hobcraft, Menken, and Preston, 1985; Keyes and Li, 2012; Yang and Land, 2013). An understanding of these concepts is foundational for determining whether workforce attitudes, values, and behaviors are attributable to generational differences, developmental differences between younger and older workers, or broad historical or social forces that impact all workers regardless of their age and generation.

In much of the academic work in these fields, age, period, and cohort are statistical concepts reflecting the impact of proximal causal processes on observed differences among people. Just knowing there is evidence for a cohort or age effect, in a statistical sense, would not tell researchers why any such differences are observed among workers or anything about the processes that created them. For example, researchers might study muscle mass and find an age effect. The observation of declines in muscle mass that are attributable to age does not explain how the functioning changes in terms of biological processes. A presumably large number of mechanisms

BOX 4-1
Age, Period, and Cohort Effects

Age is measured as time since birth and is a changing characteristic of individuals. An *age effect* occurs when individuals of different ages vary in the way they think, feel, and behave because of factors related to their stage of the life course. Age effects are considered developmental influences because they are a result of biological factors or maturation that occurs to all people regardless of when in history they were born and current historical conditions. For example, younger workers may be physically stronger (on average) than older workers because of age-related changes in muscle fibers.

Period is typically captured as the year of observation and is a changing characteristic of the broader sociohistorical context. A (time) *period effect* occurs when individuals change in the way they think, feel, and behave because of the events or social phenomena of a specific point in history. For example, the impact of a global pandemic might lead to increased anxiety for all people in a society at a given point in time, regardless of age group. After the pandemic had ended, everyone might express more apprehension about disease, even if at different levels, than they would have before the pandemic.

A cohort is a group of individuals with distinct characteristics or experiences. Cohorts are often defined as those individuals born in the same year and expected or known to have moved through their lives in concert and experienced major events at the same point in their development. The same idea applies to people who were born within a narrow set of birth years, which is why generational research often combines multiple birth years. Birth year is a fixed attribute of individuals. If a strong *cohort effect* is observed in statistical analysis, this would indicate, for example, that workers born in 1972 are categorically different from workers born in 1992 as a result of the differential influence of cultural, historical, and social events. A cohort effect differs from a period effect in that with a cohort effect, particular historical experiences influence a specific group of people because of their stage of development (or other unique characteristic) at the time of exposure, whereas a period effect impacts all people regardless of age. A cohort effect is unique to people born in a particular year or set of years because of when in their development they were exposed to particular events. For example, the events of an economic depression might make all people sensitive to financial losses after the depression (a period effect), or it might uniquely affect a group in their formative years (a cohort effect) because of the more negative emotional and economic impact on their earning potential at a time when they were entering or exploring the labor market.

For most studies of people and the variations among individuals over time, some aspects of age, period, and cohort all may contribute to the outcomes observed. The challenge for researchers is to identify which is the predominant influence.

NOTE: The content of this box is repeated from Chapter 1 for the reader's convenience.

might be relevant. The point is that the statistical modeling does not provide clear guidance as to why and how muscle mass changes with age—it simply identifies an association between age and muscle mass differences. This is an important perspective to keep in mind when considering the literature that attempts to isolate age, period, and cohort effects and interpreting results from particular studies. Theoretical knowledge and insight into mechanisms are critical. As discussed above, however, little theoretical work has examined the mechanisms that could be responsible for differences among generations. Further, it turns out to be a challenging task even to separate cohort effects from age and period effects.

It is helpful to begin with an example to illustrate the challenges of separating out these effects. Table 4-1 shows the ages of individuals of different birth cohorts at different periods in time (i.e., year of observation). The ages shown in this table are restricted to those that might be observed commonly by workplace managers, such as individuals between the ages of 20 and 70. Careful inspection of Table 4-1 gives a sense of the relations among age, period, and cohort. For example, the 20-year-olds in 2010 were born in 1990, whereas the 40-year-olds in that year were born in 1970. The table illustrates that managers in the 2010 workplace would not have observed workers in the 2000 birth cohort given that such people would be only 10 years old. And forecasting whether any generational differences

TABLE 4-1 Observed Ages of Different Workplace Cohorts at Different Periods

Birth Year	Period (Current Year)					
	2020	2010	2000	1990	1980	1970
2000	20	NA	NA	Impossible	Impossible	Impossible
1990	30	20	NA	NA	Impossible	Impossible
1980	40	30	20	NA	NA	Impossible
1970	50	40	30	20	NA	NA
1960	60	50	40	30	20	NA
1950	70	60	50	40	30	20
1940	NA	70	60	50	40	30
1930	NA	NA	70	60	50	40
1920	NA	NA	NA	70	60	50
1910	NA	NA	NA	NA	70	60

NOTES: NA = not applicable. For illustrative purposes, the table assumes that individuals below age 20 or above age 70 will not be observed in the workforce (NA). "Impossible" refers to the fact that it would be impossible to observe someone of that birth year at a given period.
SOURCE: Generated by the committee.

will persist into the future is hindered by the impossibility of a manager in 1990 having observed a member of the 2000 birth cohort. In short, there is a limit to how much cohort-related variation can be observed, and this holds true for both managers and social scientists.

Table 4-1 also illustrates that researchers taking a cross-section of workers from any given year will immediately face a problem in drawing sound inferences (see the yellow-highlighted column in Table 4-1): members of a given cohort will differ from members of other cohorts not only by cohort but also by age. Thus, any differences among these workers could be due to the effects of either age or cohort; at any single point in time (or in any period), age and cohort are confounded, so there is no way to separate the influence of the two in such cross-sectional comparisons. Researchers cannot know whether the 20-year-olds are different from the 40-year-olds because of where they are in their life span (an age effect) or the unique experiences of their generation (a cohort effect). In other words, the differences found in cross-sectional studies are consistent with either an age or a cohort effect, and nothing in the design constrains the inferences any further. Many of the studies reviewed for this report use a cross-sectional design, and as such, they offer insufficient internal validity (i.e., the study design makes it impossible to eliminate alternative explanations for any findings) to answer questions about generation effects. Such studies are therefore of limited value when thinking about the utility of generational distinctions.

One way to approach this limitation of cross-sectional studies is to compare responses from a sample of 40-year-old participants taken at different points in time. For example, if data were available on the commitment of workers to their organization in 1990 and 2020, it might be possible to compare the 40-year-olds in 2020 with the 40-year-olds in 1990 (see the orange-highlighted cells in Table 4-1). This approach holds age constant. Here again, however, a problem arises: because the workers from the different birth cohorts are observed at different times, any differences between these workers could be due to the effects of period rather than generation. It is possible that all workers observed in 2020 are less committed to their jobs compared with workers in 1990.

This discussion demonstrates that researchers attempting to separate out age, period, and cohort effects must struggle with what is known in the literature as the "identification problem"—the linear relationship among age, period, and cohort (age = period − cohort, where age is years since birth, period is current year, and cohort is birth year [see Fosse and Winship, 2019]). The identification problem makes it challenging to design a study that can distinguish cohort from age effects or cohort from period effects without making certain assumptions. In general, researchers assume that one of the three factors (age, period, or cohort) has a roughly zero effect. For example, researchers wishing to draw inferences about generations from a cross-sectional study must assume that the age effect is zero, an assumption often debated since a large body of work points to age-related differences in many of the variables of interest to those studying generations (Inceoglu, Segers, and Bartram, 2012; Kanfer and Ackerman, 2004; Kooij et al., 2011; Parry and McCarthy, 2017; Roberts, Walton, and Viechthauer, 2006).

Thus, researchers must be careful to understand the limitations of their data in the context of the research question they are attempting to answer (Blalock, 1967; Cohn, 1972; Costanza et al., 2017; Fosse and Winship, 2019). Moreover, researchers need data from multiple periods along with data on multiple ages and cohorts to even begin this process. Figure 4-1 illustrates the possible findings from a hypothetical study with enough data to distinguish age, period, and cohort effects. The next section covers some of the different designs researchers have used to study generational differences in the workforce, including the over-used cross-sectional designs and improved designs for parsing age, period, and cohort effects.

Current Analytic Approaches

As discussed above, the vast majority of studies have sought to identify generational differences using cross-sectional designs. Other quantitative methods used in this research have included cross-temporal meta-analyses and complex multilevel statistical models applied to nested datasets (see Appendix A for details). These approaches entail different ways of attempting to isolate age, period, and cohort effects, with varying degrees of success (see also Table 4-2).

Cross-sectional surveys—the approach of comparing groups of people of different ages using an instrument administered to a single sample at a single point in time. As discussed previously, such a design confounds age and cohort effects. Period effects are undetectable because all groups are completing the survey at the same time, and period effects are therefore constant.

Cross-temporal meta-analyses—the approach of extracting descriptive statistics (often measures of central tendency, such as sample means) from studies conducted at different points in time. These descriptive statistics are combined using meta-analytic techniques and usually weighted for precision by the number of observations available for each time point. The objective is to test whether aggregated estimates vary because of when the data were collected. If the underlying studies used age-restricted samples (e.g., samples of high school students, college students, or new Army recruits), it is common for the results to be used to draw inferences about generations. As previously discussed, however, an observation of

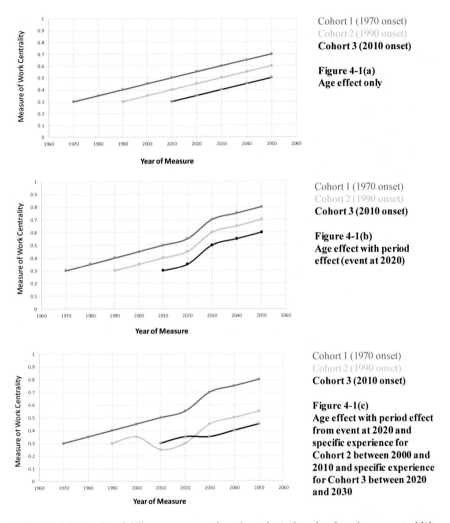

FIGURE 4-1 Graphs of different outcomes for a hypothetical study of work as a central life interest over time.

NOTES: In this hypothetical study, the value of work centrality (how important work is in one's life) is measured every 10 years using the same validated instrument with three different cohorts or samples of adults, starting when they are age 20. This hypothetical example assumes very little variance within samples at a given time. In (a), results in straight lines of the same slope would indicate that the value of work centrality is strongly dependent on age. In (b), a significant event occurring in 2020 affects all in ways that increase the value of work. The resulting graph illustrates a period effect that would perturb the age relationship. In (c), the age relationship is perturbed by a period effect from an event occurring in 2020, as well as a significant experience only for Cohort 2 from 2000 to 2010 and a significant experience only for Cohort 3 from 2020 to 2030. The resulting graph illustrates two different cohort (or generation) effects that are as significant as the age and period effects on the samples.

TABLE 4-2 Overview of the Analytic Approaches

Analytic Approach	Data Requirements	Advantages and Disadvantages
Cross-sectional surveys (e.g., see list of studies in Appendix A)	Data from a survey administered to a single sample across multiple ages or generations at a single time point, analyzed and summarized statistically (means, standard deviations) for each age group	Period effects are held constant, but cohort and age are confounded
Cross-temporal meta-analyses (e.g., Campbell, Twenge, and Campbell, 2017; Twenge, Campbell, and Freeman, 2012; Twenge and Campbell, 2001, 2008)	Descriptive statistics (sample means, standard deviations) from studies of people of the same age sampled at different time points	Controls for age effects, but period and cohort are confounded
Multilevel models (e.g., Donnelly et al., 2016; Jürges, 2003; Kalleberg and Marsden, 2019; Koning and Raterink, 2013; Kowske, Rasche, and Wiley, 2010)	Individual-level data collected from multiple survey panels repeatedly over extended periods of time	Partitions the variance attributable to age, period, and cohort (generation); relatively few available datasets have information relevant to workplace considerations; statistical assumptions are often complex and untested

SOURCE: Generated by the committee with information from Costanza et al., 2017.

differences across different years could be attributable to period rather than cohort effects. For example, researchers might identify all studies that administered the same measure of organizational commitment across different years. The sample means for organizational commitment would be recorded from each study, along with the year of data collection. The sample means would be combined for each year to generate a more precise estimate of organizational commitment by pooling results from multiple studies. Researchers could then test whether the pooled averages for each year showed any systematic fluctuations across different years. Such a systematic pattern would suggest changes in organizational commitment over time; however, it would be impossible to discern whether any observed changes were due to period or generation effects.

Multilevel models applied to nested datasets—Multilevel models are a family of statistical tools that are appropriate for studying databases in which some observations are nested within others, such as when data are collected from multiple individuals across multiple years. Statistically speaking, individual responses are then nested within each year in this design, which is essentially a series of repeated cross-sectional studies with data being collected across multiple years. Likewise, nesting can occur in longitudinal designs when the same people are observed repeatedly over time, as is typically the case in many disciplines. In this case, observations on different occasions are nested within people. Still other cases of nesting arise when observations are clustered within groups, such as students within schools or employees within workplaces. All of these kinds of designs raise the possibility that observations are not completely independent from one another because of the nesting. Multilevel models are statistical tools that allow researchers to address the nesting.

The statistical approach of using multilevel models to study generation effects is technical (see Yang and Land, 2013) and not without critics (see Bell and Jones, 2018). Concerns involve the statistical assumptions, coding of the data, and modeling constraints needed to achieve estimation of the model given the identification problem described in the previous section and the nature of the data that contain multiple ages observed across multiple time points. In the generational literature, such multilevel models are called age-period-cohort (APC) models or APC analysis (Fannon and Nielsen, 2019; Fosse and Winship, 2019; Winship and Harding, 2008). APC models are applied to datasets that include multiple ages and times of measurement. These datasets need not necessarily be longitudinal in the sense of following the same individuals across multiple time points. Instead, different people can provide information at different waves. For example, data from the General Social Survey (GSS) (https://gss.norc.org) are often used for APC analyses. The GSS has administered the same questions (e.g., "Taken all together, how would you says things are these days—would you say that you are very happy, pretty happy, or not too happy?") across multiple years (e.g., from 1972 to 2018) to people of varying ages at each year (i.e., people between the ages of 18 and 89 or older). Data from the GSS can be analyzed with multilevel models to isolate how much variability in a given variable is attributable to the statistical effects of age, period, and cohort. The issue with this approach is that relatively few datasets with information relevant to workplace considerations are available for analysis. Moreover, as noted above, the modeling is often not without assumptions that could be challenged on both statistical and theoretical grounds (e.g., Bell and Jones, 2018; Luo et al., 2016).

Other Methodological Concerns

As discussed in the previous section, cross-sectional surveys are typically not useful for studying generational differences because they confound age and cohort effects, and one of the most promising approaches for addressing this limitation is to use APC statistical models on repeated datasets that span many years to provide multiple observations of people of different ages at different points in time. However, issues of measurement invariance and representativeness are also relevant when evaluating the literature on generational differences. These issues are salient to the study of generations because it is important to confirm that the tools (e.g., surveys) used to measure constructs (e.g., work-related values) of interest are able to support comparisons across the targeted generational groups (measurement invariance) and whether population-level inferences are justified given the sampling plan (representativeness).

Measurement Invariance

Researchers interested in workplace characteristics typically focus on such topics as job satisfaction, intrinsic motivation, and organizational commitment, termed "constructs" in the social sciences. Most social scientists acknowledge, either implicitly or explicitly, that survey responses involve some degree of imprecision. The actual responses are reflections of the underlying construct of interest; they are not thought to be perfect indicators of the construct. In fact, questions about reliability and validity typically arise for all measures. Rigorous examinations of generational differences therefore require that survey responses (or any other measures) operate in the same way across time and across members of different generations. This issue is often the domain of measurement specialists (i.e., psychometricians) who are concerned with psychometric properties and the extent to which those properties change across points of comparison. Other social scientists might simply assume that measures have the same psychometric properties. Nonetheless, measurement equivalence/invariance (e.g., Horn and McArdle, 1992) is critical for drawing the appropriate conclusions from quantitative data in the generational literature.

The basic question with respect to measurement invariance is whether psychometric properties are consistent across groups or time points so that the observed scores reflect the same value of the construct (also known as a latent variable in some disciplines) whenever comparisons are made. If this condition of invariance is not met, sound inferences across studies are impossible. For example, consider a study comparing average scores on a multi-item survey of job satisfaction for younger employees (i.e., workers under 30) versus older employees (workers over 55). The measure of job satisfaction would be considered invariant if the observed scores referred to the same underlying level of job satisfaction for both groups of workers (i.e., if a score of 3.5 referred to the same level of satisfaction for a 25-year-old and a 55-year-old). If such a condition held, it would then be reasonable to draw inferences about observed differences in average levels of job satisfaction between the older and younger workers. If measurement invariance did not hold, however, the same observed scores would refer to different levels of job satisfaction in the two age groups. If invariance is not present, drawing inferences from comparison between groups is akin to the proverbial problem of comparing apples and oranges (Vandenberg and Lance, 2000).

There are different levels of measurement invariance (Schmitt and Kuljanin, 2008; Vandenberg and Lance, 2000), imposing increasingly stringent requirements on the psychometric properties of scores. Invariance is evaluated using structural equation modeling techniques (e.g., Brown, 2015) or item-response theory methods (Tay, Meade, and Cao, 2015). Both of these techniques are predicated on the notion that observed scores on quantitative measures reflect differences in underlying latent (unobserved) variables. Both techniques formally acknowledge measurement imprecision and do not assume that observed scores are perfect reflections of the constructs they are intended to measure.

Many studies in the generational literature fail to test explicitly for measurement invariance, adding further ambiguity to attempts to draw conclusions from the existing literature. Only two studies included in the committee's review directly examine measurement invariance in the context of generational differences (Meriac, Woehr, and Banister, 2010; Twenge et al., 2010). These studies show work values to be partially invariant across three generational groups and thus support the idea that cross-generational comparisons are meaningful. Inferences from these studies are based on much stronger psychometric ground with respect to making comparisons; however, the approaches taken are still constrained by the potential confounding of age, period, and cohort effects.

Representativeness

In addition to measurement consistency, it is important to consider the nature of the samples used for generational research. Inferences about populations are only as sound as the sampling strategy of a given study. Rigorous approaches to the selection of sample subjects can strengthen the external validity of studies (or the degree to which inferences from a study can be extended to larger populations of interest). The issue of representativeness is critical to external validity. In the case of generational research, the relevant issues are whether the samples are truly representative of the generations of interest and whether the diversity of the population is represented in the samples. Given demographic shifts and changes, comparing samples of workers from the 1980s to the 2020s involves comparing samples that vary in terms of many characteristics, such as ethnicity, race, parental education, and income. It is important for the existence of demographic differences and diversity to factor into the interpretations of differences among generations. Put simply, researchers need to consider how demographic differences may confound generational comparisons.

Although researchers sometimes use the phrase "representative" to apply to samples drawn from defined populations, such a term is probably best applied to the process used to generate a given sample (see Stuart, 1968, as cited in Pedhazur and Schmelkin, 1991). In classic survey methodology, samples are drawn from defined populations of interest because, while researchers are interested in drawing conclusions about a population, they often lack the time, money, or other resources to collect information from every element or member of that population. In fact, the use of inferential statistics eliminates the need to study all members of a population.

The issue with representativeness becomes how well the samples ultimately generated by researchers are representative of the population of interest. A convenience sampling strategy uses no randomization, simply taking advantage of accessible members of a population (e.g., employees willing to fill out a survey, college students enrolled in introductory psychology courses); therefore, this strategy can lead to sampling a biased subset of the population. Because there is no formal way to estimate sampling errors in convenience samples (Pedhazur and Schmelkin, 1991), it is impossible to estimate how well the characteristics of such samples reflect the attributes of the population of interest. In lieu of drawing on formal statistical principles, then, researchers must make educated guesses. The bottom line is that representativeness is unknown and unknowable when convenience samples are used.

Probability samples are usually more difficult to collect and require that researchers determine the odds that any element of a population will be selected for inclusion in the sample. The simplest case is when all elements have the same nonzero probability of being selected. There are, however, complicated sampling strategies involving stratification and over- and undersampling, which are widely used in polling applications and epidemiology. The virtue of these probability-based methods is that sampling errors can be calculated, placing inferences drawn from extrapolating the sample to the population on much stronger footing. Still, nonresponses can bias a sample if the chances of not participating in a sample (i.e., by refusing to consent or being unable to complete a survey) differ across different subsets of the population. Thus, even in ideal cases in which organizations use scientific sampling, questions about representativeness can remain.

The complexities of probability sampling and survey nonresponse are largely beyond the scope of this report. However, these issues are relevant to the analysis of the strength of the evidence for generational differences. Consumers of generational research thus need to evaluate whether issues of sampling and representativeness are approached in thoughtful ways. For example, how well do samples represent the generations of interest? Does the literature consider other sample characteristics (e.g., sex, race/ethnicity, education level) or just birth years, given that such characteristics may moderate or be alternative explanations for observed effects? When researchers ignore sampling issues, deficiencies in the rigor of the work are likely.

Qualitative Studies

Although most of the empirical studies reviewed in this report are quantitative, there are also a number of relevant qualitative studies (see Appendix A). The qualitative approaches entail analyzing data in the form of natural language (i.e., words) and expressions of experiences (e.g., social interactions). The various methods[3] differ in representing a diversity of philosophical assumptions, intellectual disciplines, procedures, and goals (Gergen, 2014). Nevertheless, these methods all share an iterative process of evolving findings (e.g., as driven by induction) and viewing subjective descriptions of experiences as legitimate data for analysis (Wertz, 2014).

Using an iterative process to draw inferences means that researchers tend to analyze data by identifying patterns tied to instances of a phenomenon and then developing a sense of the whole phenomenon as informed by those patterns. Seeing the patterns can shift the way the whole is understood, just as seeing a pattern in the context of a whole phenomenon can shift the way the pattern is understood (Levitt et al., 2018). These iterations are self-correcting; as new data are analyzed, the analysis corrects and refines the existing findings.

Among the 29 studies reviewed by the committee that use qualitative approaches to assess generational characteristics, 15 explicitly compare different generation groups. The sampling methods include purposive and convenience sampling. With purposive sampling, the sample is chosen for purposes of maximizing variability, generating typical or critical cases, covering extreme/deviant situations, and gauging expert opinions. The sample sizes range from single digits (e.g., in case studies) to more than 100 (e.g., in larger interview studies and discursive analyses). When convenience sampling is used, researchers often explicitly justify the legitimacy of that sampling in their specific research context. When researchers have the goal of identifying generation differences in certain domains, they often sample based on the birth cohort categorizations of various generations. Neverthe-

---

[3] A range of qualitative analytic approaches—such as narrative, grounded theory, phenomenological, critical, discursive, case study, and thematic analysis approaches—have been used in the generational literature (Lichtman, 2014).

less, as discussed above, it is clear that such sampling cannot separate age effects from the intended generation effects.

In the generational literature, researchers typically use qualitative methods to answer two broad research questions: (1) Do generational differences *exist* in certain attributes, behaviors, attitudes, or values? and (2) Do people *perceive* generational differences in certain attributes, behaviors, attitudes, or values? The main qualitative data collection method used in the literature to address the first of these questions is interview. Assuming the interviewees represent the intended generational groups (either through interviewees' self-identification or through arbitrary categorization based on the span of birth years), researchers derive the attributes, behaviors, attitudes, or values of interest from the interview responses and compare them across the intended generational groups. In addition to the potential methodological issues involved in analyzing interview responses (e.g., interpretation bias, coding unreliability), an obvious issue with this approach is that neither self-identification with the targeted generational groups nor arbitrary categorization based on the span of birth years can rule out the confounding effects of age and period discussed earlier. Accordingly, even if systematic differences are seen in interview responses across the intended generational groups, it is unclear whether those differences are due to generation, age, or period effects. This methodological weakness due to grouping applies as well to other qualitative data collection methods (e.g., observation, focus group discussion, document analysis). Therefore, qualitative methods do not offer sufficient internal validity in addressing the first research question above.

The main qualitative data collection methods used to address the second research question include interview, focus group discussion, and document analysis. Given the focus of this research question on the *perception* of generational differences, participants in interviews and focus group discussions need to be made aware of the concept of generations before they report the differences they perceive. The issue here is that the generation-related information given to participants may influence how they retrieve their memories and experiences or form their impressions and judgments. When document analysis is used, this issue is of less concern because the document content is typically archival in nature and is generated independently from the research purpose. Regardless of the qualitative methods used, however, if the sampling coverage is narrow, any findings about people's perceptions of generations cannot be generalized to all people and may therefore better be treated as preliminary and used to inform subsequent quantitative investigations.

To facilitate communication of the findings obtained with qualitative methods, it is best practice for researchers to describe the origins or evolution of their data collection protocol so that other researchers can assess how the concept of generations was introduced to study participants and make judgments about interpretations of the findings. Further, researchers are advised to explicate in detail the process used for analysis, including some discussion of the procedures involved (e.g., coding, thematic analysis), adhering to the principle of transparency (Levitt et al., 2018). This discussion also would include describing coders or analysts and their training, as well as what software was used for the data analysis. It is important to identify clearly whether coding categories emerged from the analysis or were developed a priori. Triangulation across multiple sources of information, findings, or investigators is typically viewed as desirable in terms of generating strong support for the research claims. However, the committee found very little application of these best practices in the qualitative studies in the generational literature.

Qualitative methods can be used to achieve such research goals as developing theory, hypotheses, and attuned understandings; examining the development of a social construct; and illuminating social discursive practices (i.e., the way interpersonal and public communications are enacted) (Levitt et al., 2018). It is the committee's belief that in research on worker attitudes and behaviors, the continued use of qualitative methods is to be encouraged. Because of the limitations discussed throughout this chapter, qualitative studies cannot verify the existence of generational differences. When appropriately designed and documented, however, they can help advance understanding, for example, of such work-related constructs as job satisfaction, as well as of generational perceptions that affect workplace behaviors. (See the further discussion of alternative perspectives for future research in Chapter 5.)

Mixed Methods

According to Creswell (2015), the use of mixed methods involves (1) collecting and analyzing both qualitative and quantitative data in response to overarching research aims, questions, and hypotheses; (2) using rigorous methods for both qualitative and quantitative research; (3) integrating or "mixing" the two forms of data intentionally to generate new insights; (4) framing the methodology with distinct forms of research designs or procedures; and (5) using philosophical assumptions or theoretical models to inform the designs. The committee's review of the generational literature revealed six studies employing both quantitative and qualitative methods. However, these studies appear to have the same weaknesses identified above—insufficient internal and external validity in both the qualitative and quantitative portions to justify inferences about generational differences. Although rarely used appropriately in generational research, however, mixed-methods approaches could lead to additional insights not gleaned from qualitative or quantitative findings alone (Creswell, 2015). The value of using mixed methods accrues from the integration of qualitative and quantitative findings in a thoughtful way that leads to greater mining of the data and enhanced insights. In principle, the use of mixed methods has the potential to lend credibility and robustness to research designs.

DISCUSSION

Since the late 1990s, the number of empirical studies on generational differences in work values/attitudes has increased dramatically. These studies generally use birth cohorts to define generations and draw on popular labels to categorize groups in their samples. While popular notions of generations have become broadly familiar, the wide range of birth years used to identify various generational groups indicates a lack of consensus on how generations should be operationalized in research.

Most generational researchers have approached the empirical study of generational differences with the underlying assumption that the overall concept of "generations" is valid. They take at face value that an undefined set of shared experiences—social, political, cultural, and historical influences—have shaped the attitudes/values to be measured. To date, however, little theoretical or empirical justification has been offered to clarify the events and shared experiences assumed to define a generation. At best, the work usually is purely descriptive.

This research has been motivated by a desire to understand generational shifts in the workforce and their impacts on such employment practices as recruitment, retention, and training. While this is a worthwhile research pursuit, the existing generational literature has a number of limitations: (1) untested assumptions and conceptual variations regarding the concept of generation; (2) overreliance on cross-sectional studies and convenience samples, which have relatively weak internal and external validity with regard to the objectives of identifying generational differences and generalizing findings to all members of each generation; and (3) statistical challenges in separating out age, period, and cohort effects, even with the more rigorous research designs. Together, these limitations call into question whether researchers can draw sound inferences from the existing literature.

Conclusion 4-2: The body of research on generations and generational differences in the workforce has grown considerably in the past 20 years. Despite this growth, much of the literature suffers from a mismatch between a study's objectives and its research design and underlying data, which threatens both the internal and external validity of the work. The research designs and data sources rely too heavily on cross-sectional surveys and convenience samples, which limits the applicability and generalizability of findings.

While much of the literature reviewed by the committee relies on one-time, cross-sectional surveys whose results confound age and cohort effects, some researchers have used multilevel models, discussed above, to distinguish cohort effects from age and period effects (e.g., Donnelly et al., 2016; Jürges, 2003; Kalleberg and Marsden, 2019; Koning and Raterink, 2013; Kowske, Rasche, and Wiley, 2010; Leuty and Hansen, 2014; Lippmann, 2008). Kalleberg and Marsden (2019) illustrate such a statistical approach to disentangling the effects of age, historical time period, and generation (i.e., cohort differences) on changes in work values in the United States. They use data from the GSS (1973–2016) and the International Social Survey Program (ISSP) (1989, 1998, 2006, 2016). These datasets consist of information collected from multiple cross-sectional samples designed to represent the U.S. population in the various years and thus provide repeated value measurements across ages and time. The authors analyze these data using hierarchical logistic regression analyses in which period and cohort differences are modeled using random effects (i.e., a multilevel model applied to repeated surveys administered across multiple years). Work values are conceptualized in two different ways given the data available in the two surveys. The first (measured in the GSS datasets) involves work as a "central life interest," with respondents being asked whether they would continue to work or stop working if they were wealthy enough to have that option. The second (measured in the ISSP datasets) entails asking respondents to rate the importance to them of different features of jobs (which are measured as single items) (ratings range from "not at all important" to "very important"). The job features measured are both extrinsic (security, high income, potential for advancement) and intrinsic (interesting work, opportunity to help others, opportunity to help society), as well as flexible hours.

Kalleberg and Marsden (2019) find little evidence for pronounced generational (i.e., cohort) differences in work values. While these differences may be statistically detectable, they are substantively minor. This finding suggests that much speculation about the distinctiveness of values—such as being self-absorbed and narcissistic (Twenge, 2006) or less concerned with career advancement than with achieving greater work–life balance (Jenkins, 2018) for particular generations lacks a strong empirical grounding, at least for the United States. Rather, these authors found that age differences were dominant in explaining differences in whether respondents would continue to work if they were wealthy enough not to have to do so. The idea that work is a central life interest declined by age until age 65, after which it increased somewhat. On the other hand, variations in the time periods during which people live are most closely related to changes in the importance they place on the various facets of jobs. Thus since the 1990s, people in the United States have tended to place greater importance on jobs that provide security, high income, and more opportunities for advancement. These patterns are consistent with the view that these job features have become more difficult for workers to attain in recent years.

The authors of many studies that claim to support generational differences could not disentangle whether age differences, changes between time periods, or distinctions between generations were the root cause of observed effects. Because of inherent challenges in studying cohort or generation effects, many researchers may have misattributed their own findings or the findings of others to generational differences. In so doing, researchers themselves have helped precipitate the conclusion that younger generations of workers are somehow different from previous generations. For instance, the analysis of differences in work values by Twenge and colleagues (2010) is commonly interpreted as providing evidence for generational differences. Yet their analysis was limited to a comparison of 16-year-olds in three different decades. Consequently, although the study provides evidence for time-related differences in work values, it is possible that had older individuals also been sampled over time, the researchers might have observed the same changes in the older group. This observation would have indicated that changes between time periods, not generational differences, better explain the observed differences. In fact, these authors acknowledge the ambiguity in their results and point to period effects as an alternative explanation for their findings.

As reviewed above, a small subset of studies have used APC methods to examine work-related attitudes and values. The study by Kalleberg and Marsden (2019) provides analyses of some of the very work values reviewed in the aforementioned study by Twenge and colleagues (2010). However, when these authors used APC analyses, they found that observed changes were not a function of generational differences at all; instead, period effects were at the root of the changes. The contrast between these two studies is telling, showing that when more rigorous methods are used, what appears to be attributable to generation effects can actually be attributable to period effects. Unfortunately, very few studies examining worker attitudes and values have used APC methods.

Many more studies have used APC methods to disentangle these effects in other domains (e.g., in the examination of changes in self-esteem over time by Twenge and colleagues (2017). These studies typically find that when time-based changes are analyzed, the period effects are much greater than the generation effects, and when generation effects are present, they tend to be small. Given these findings, the use of APC models in future research examining changes in work-related variables is the best way to offer less unambiguous conclusions. However, use of this approach may be constrained by issues with data availability. Likewise, it is important for researchers to specify carefully the statistical assumptions behind the multilevel model and to evaluate critically whether they are tenable. With respect to cross-temporal meta-analysis, this approach is imperfect in that it does not allow for the separation of period and cohort effects. However, it is a useful tool for determining whether a given construct has changed over time in general, and research examining psychological variables using cross-temporal meta-analysis continues to be useful for that purpose. Finally, cross-sectional studies with convenience samples have limited utility, and their findings cannot be used appropriately if the goal is to draw inferences about generational differences.

> **Recommendation 4-1:** Researchers interested in examining age-related, period-related, or cohort-related differences in workforce attitudes and behaviors should take steps to improve the rigor of their research designs and the interpretation of their findings. Such steps would include
>
> - decreased use of cross-sectional designs with convenience samples;
> - increased recognition of the fundamental challenges of separating age, period, and cohort effects;
> - increased use of sophisticated approaches to separate age, period, and cohort effects while recognizing any constraints on the inferences that can be drawn from the results;
> - greater attention to the use of samples that are representative of the target populations of interest;
> - greater attention to the design of instruments (e.g., surveys) to ensure that the constructs of interest (i.e., measured attitudes and behaviors) have the same psychometric properties across time and age groups; and
> - increased use of qualitative approaches with appropriate attention to documenting data collection protocols and analysis processes.

5

Alternative Perspectives for Research

As discussed in Chapter 4, given the conceptual and methodological limitations entailed in separating out variations among individuals due to effects of age, generation (birth cohort), or social change (period), there is currently no strong empirical evidence for shared differences that would distinguish one whole generation from another. Chapter 4 describes these limitations and calls for improved attention to methods used, internal and external validity, and the conclusions that can appropriately be drawn from any findings in future research examining age-related, period-related, or cohort-related differences in workforce attitudes and behaviors.

Despite the lack of scientific evidence for generational differences, an interesting reality is that people tend to believe these differences exist, a belief that sustains perceptions of many differences among workers (Lester et al., 2012; North and Shakeri, 2019). Therefore, it could be more fruitful for future research to focus on examining *beliefs* and *perceptions* about the qualities possessed by generations and their impacts in the workplace (Costanza and Finkelstein, 2017; Weiss and Perry, 2020), keeping in mind that these beliefs and perceptions may not reflect true attributes of any birth cohorts and thus can be studied as generational stereotypes and biases (Perry et al., 2017). For example, it is important to understand how beliefs and perceptions about generational attributes develop. Research could examine how individuals develop beliefs and perceptions about their own birth cohort, as well as about the generations/birth cohorts of others. It is also important to understand how beliefs and perceptions about generations impact behaviors and interactions in the workplace. Indeed, if those impacts occur mainly through stereotypes and biases, understanding of these impacts can have significant implications for managing workers fairly.

To help inform future research, this chapter examines (1) the inherent appeal of and major psychological motivations for using generational ascriptions, (2) the risks of using those ascriptions in the workplace, and (3) additional perspectives on the multiple influences on workforce development over time.

THE INHERENT APPEAL OF GENERATIONS

The idea of the younger generation's being different from the older generation is an idea that goes back thousands of years. Observations made about the new generation of youth and young adults are strikingly similar throughout the ages, with the older generation complaining that individuals in the younger generation have poor morals; are degrading the language; and are lazy, thoughtless, and selfish (see Box 5-1). Thus, the idea that the steady flow of new humans can be cleaved into groups that are distinctly different from one another is not new, nor is the conflict between one generation and another. Nonetheless, the notion that each generation is decidedly less motivated or capable than the one before is not supported by either common sense or empirical evidence. If the observations about young people made repeatedly over time were true, societies would have quickly devolved instead of making steady progress in business, public health, and other areas.

Negative attitudes toward older adults have persisted for some time, with older adults often being portrayed as out of touch or incompetent. For example, in 2019, the phrase "OK boomer" exploded across social media and even spread as far as the New Zealand Parliament (Mezzofiore, 2019) and the U.S. Supreme Court (Liptak, 2020). This phrase encapsulates young people's anger at the hand they feel they have been dealt—including climate change, rising inequality, and student debt—by older generations and conveys the sentiment that older people "just don't get it." Young people use the phrase as a retort to older people who "don't like change," "don't understand new things," and "don't understand equality" (Lorenz, 2019).

Reasons for the current pervasiveness of the concept of generations, their labels, and assumed differences among them may be that they have become strongly socially constructed over time and have emerged to serve a purpose in social identities (Rudolph and Zacher, 2015, 2017). Humans are naturally inclined to categorize and generalize, skills that are useful in deciding quickly whether a situation is dangerous or simplifying a mass of information that one needs to process (Rosch, 1978). Social categorization is

---

BOX 5-1
Quotations about the New Generation

4th century BC: "[Young people] are high-minded because they have not yet been humbled by life, nor have they experienced the force of circumstances.... They think they know everything, and are always quite sure about it." (Aristotle, *Rhetoric*)

20 BC: "Our sires' age was worse than our grandsires'. We, their sons, are more worthless than they; so in our turn we shall give the world a progeny yet more corrupt." (Horace, Book III of *Odes*)

1330: "Modern fashions seem to keep on growing more and more debased.... The ordinary spoken language has also steadily coarsened. People used to say 'raise the carriage shafts' or 'trim the lamp wick,' but people today say 'raise it' or 'trim it.' When they should say, 'Let the men of the palace staff stand forth!' they say, 'Torches! Let's have some light!'" (Yoshida Kenkō, *Tsurezuregusa [Essays in Idleness]*)

1771: "Whither are the manly vigor and athletic appearance of our forefathers flown? Can these be their legitimate heirs? Surely, no; a race of effeminate, self-admiring, emaciated fribbles can never have descended in a direct line from the heroes of Poitiers and Agincourt...." (Letter in *Town and Country* magazine)

1843: "...a fearful multitude of untutored savages...[boys] with dogs at their heels and other evidence of dissolute habits...[girls who] drive coal-carts, ride astride upon horses, drink, swear, fight, smoke, whistle, and care for nobody... the morals of children are tenfold worse than formerly." (Anthony Ashley Cooper Shaftesbury, 7th Earl of Shaftesbury, Speech to the House of Commons)

1904: "Never has youth been exposed to such dangers of both perversion and arrest as in our own land and day. Increasing urban life with its temptations, prematurities, sedentary occupations, and passive stimuli just when an active life is most needed, early emancipation and a lessening sense for both duty and discipline, the haste to know and do all befitting man's estate before its time, the mad rush for sudden wealth and the reckless fashions set by its gilded youth— all these lack some of the regulatives they still have in older lands with more conservative conditions." (Granville Stanley Hall, *The Psychology of Adolescence*, pp. xv-xvi)

1925: "We defy anyone who goes about with his eyes open to deny that there is, as never before, an attitude on the part of young folk which is best described as grossly thoughtless, rude, and utterly selfish." ("The Conduct of Young People," *Hull Daily Mail*, September 11, p. 4)

1936: "Probably there is no period in history in which young people have given such emphatic utterance to a tendency to reject that which is old and to wish for that which is new." ("Young People Drinking More," *Portsmouth Evening News*, October 6, p. 9)

1993: "What really distinguishes this generation from those before it is that it's the first generation in American history to live so well and complain so bitterly about it." ("The Boring Twenties," *Washington Post*, September 12)

SOURCES: Ferry (2015); Georges (1993); Hall (1904); Ogborn (1998); Ross (2010); Shaftesbury (1868); and Yoshida and Keene (1998).

a cognitive process by which individuals place other people and themselves into social groups (Allport, 1954/1979; Fiske and Taylor, 1991). This categorization can be influenced by various sociocontextual elements, including existing labels and definitions (Brewer and Feinstein, 1999; Fiske 1998). Social categorization is common for observable characteristics, such as when people encounter, and draw quick inferences about, individuals of a certain race, gender, linguistic group, or age (Liberman, Woodward, and Kinzler, 2017). Often linked to impressions of age, generational categories provide a way to quickly stereotype other people's values, skills, and tendencies based on ideas about their generation as a whole (Costanza and Finkelstein, 2015).

The existing literature on the formation of beliefs and perceptions about generations is limited. Some research has shown that as people grow older, they develop a more positive view of their birth cohort as compared with their age group (Weiss and Lang, 2009). One explanation is that in contrast to age identity, which is often perceived as threatening, generation identity represents a resource in later adulthood that provides a sense of agency, positive self-regard, and continuity (Weiss and Lang, 2012). Supporting this explanation, Weiss and Perry (2020) show that generational metastereotypes compared with age metastereotypes (i.e., what people think other people believe about their generation or age group, respectively) positively influence older adults' work-related self-concept. This line of research attempts to explain how people develop beliefs and perceptions about their own generation/birth cohort, while other research is focused on understanding how people develop beliefs and perceptions about others' generations/birth cohorts.

Protzco and Schooler (2019) attempted to study systematically whether and why people believe that today's youth are deficient relative to previous generations. They asked the study participants to rate the children of today against the children of their own generation on a variety of traits. The researchers also assessed the participants themselves on those same traits, including respect for authority, intelligence, and enjoyment of reading. They found that people who excelled on a particular trait in question were more likely to believe that children are in decline on that same trait. Compared with participants who do not like to read, for example, participants who like to read were more likely to believe that children these days like to read less compared with children when the participants were young. The researchers theorized that denigrating today's youth is a fundamental illusion grounded in several cognitive mechanisms, including a bias toward seeing others as lacking in traits in which one excels and a memory bias that projects one's current traits onto past generations. To denote this phenomenon, Protzco and Schooler coined the term the "kids these days effect," noting that complaints about "kids these days" have been pervasive through millennia of human existence.

RISKS OF USING GENERATIONAL CATEGORIES

Generational categories have taken on a life of their own (see Chapter 3). The labels used to describe people can shape the way they are perceived, regardless of whether those labels are accurate (Darley and Gross, 1983). While social categorization can be useful in some instances, it can also lead to prejudice, bias, and inappropriate stereotyping (Liberman, Woodward, and Kinzler, 2017). In this way, categorizing and labeling people—whether on the basis of generation, race, gender, sexual orientation, religion, or some other characteristic—can be dangerous and harmful. In the present context, the use of one label for everyone born within a particular time period can lead to stereotyping with generalities that may or may not be true for particular individuals. People born in the same year or span of years may have very different experiences, depending on such factors as socioeconomic status, geographic location, education level, gender, and race. These factors modify how people experience events that are supposedly formative for their generation. For example, people who came of age during the time of the civil rights movement likely had very different experiences depending on their race and the region of the country in which they lived. Even generational labels themselves can be exclusionary and ignore the heterogeneity within generations. For example, Erica Williams Simon notes that calling the youngest generation iGen (in reference to the iPhone) would "exclude a lot of people" who lack the access to technology of higher-income people. She argues that it is "very hard to label something in a way that reflects everyone's experience" (Raphelson, 2014).

Labels that start out benign often can become pejorative over time as people emphasize their negative over their positive connotations. The *Wall Street Journal* (2017) published a note on its style blog suggesting that the term millennials had become "snide shorthand" in the paper, and observing that millennials span a wide age range and that some are leaders and shapers of society. The note asserts that many of the habits attributed to millennials are actually just common among young people in general, and that if journalists are simply referring to young people, they should "probably should just say that."

In the workplace, some studies show that people's stereotypes about different age or generation groups, or their perceptions of such stereotypes, can influence how they perform and how they interact with others, as well as drive intergenerational conflict in the workplace (Urick et al., 2017). One survey found that older and younger workers thought others viewed them more negatively than they actually did; for example, older people thought others might stereotype them as "boring" and "stubborn," whereas people actually believed older workers were "responsible" and "hard-working" (Finkelstein, Ryan, and King, 2013). Another study found that when workers believed other people held negative stereotypes about their age group, they responded either by becoming anxious and worried about how to perform, or by becoming indignant and challenging themselves to defy the stereotype (which could result in negative or positive actions) (Finkelstein et al., 2020). In a qualitative study, Urick and colleagues (2017) used interview data to identify a variety of possible tensions between younger and older workers stemming from perceptions of generation-based differences in values, behaviors, and identities. Based on these findings, it appears that interventions designed to deal with intergenerational issues at work can be more successful if they lead workers to see more similarities across generational groups (Costanza and Finkelstein, 2017). (Workforce management is discussed further in Chapter 6.)

PERSPECTIVES TO ADVANCE RESEARCH

In the absence of a compelling alternative to a generational mindset in decision making, people will likely continue to use generational heuristics to make decisions even if doing so ultimately has null, negative, or unintended consequences leading to workplace discrimination (Costanza and Finkelstein, 2015). It appears that, intuitively, people are inclined to agree with early sociological formulations according to which events that co-occur with developmentally critical periods (e.g., late adolescence/early adulthood) will influence attitudes and values in a significant way. Although compelling, however, generational paradigms tend to ignore the diversity of behavior, attitudes, and values within a generational group. Specific events occurring during critical periods of development may shape attitudes and values, but these effects appear to be influenced by life events and idiosyncratic experiences related to one's social class, geographic location, and other factors (Baltes, Reese, and Lipsitt, 1980).

The task for researchers is to identify alternatives to current theory and research on generations that are better at exploring the ways in which people's experiences—both shared and individual—may affect their work-related attitudes and behaviors. Described below are three perspectives that can be taken in thinking about the variations among workers: lifespan development theories, changes in the work context, and the aging workforce. These are not the only perspectives that may guide future research—the committee recognizes that further study and theory development may produce perspectives of better value for understanding workforce issues—but these three perspectives have existing scientific literature upon which to build.

Lifespan Development Theories

Like the sociological theories and popular generational ideas described in Chapter 3, Baltes and colleagues' lifespan development formulation considers individual characteristics and sociocultural influences on human development (Baltes, Reese, and Lipsitt, 1980). Several other scholars have also taken a lifespan approach to understanding individual development (e.g., Durham, 1991; Lerner, 2002; Li, 2003; Magnusson, 1996). Lifespan development theories differ from generational ideas in that they do not presume generational categories (Rudolph and Zacher, 2017). Rather, lifespan development theories posit three influences on identity formation—biological and environmental determinants—that provide a lens through which people interpret their experiences:

- *Normative age-graded (ontogenic) influences.* These are defined as "biological and environmental determinants that have...a fairly strong relationship with chronological age" (Baltes, Reese, and Lipsitt, 1980, p. 75). They also include sociocultural events that happen around the same age (e.g., education, marriage, parenthood).
- *Normative history-graded (evolutionary) influences.* These environmental and biological determinants are associated with historical time. They include significant events (e.g., war) and sociocultural phenomena (e.g., social media) that are normative in that they affect everyone who experiences them.
- *Non-normative life events.* These biological and environmental determinants are idiosyncratic to individuals. Examples are person-specific personality traits and abilities, knowledge acquired through individual experiences, individual illnesses that may cause an impairment, and opportunities influenced by a person's family station and socioeconomic status.

Lifespan development theories posit that each of these influences has a different trajectory (Baltes, Reese, and Lipsitt, 1980); see Figure 5-1. Normative age-graded influences have a greater effect in early and late developmental periods, when people purportedly have less agency, and can be represented with a u-shaped curve. Normative historical events have a greater effect during critical developmental periods (e.g., late adolescence/early adulthood) and can be represented with an inverted u-shaped curve. And the influence of non-normative life events becomes increasingly important through the life span and can be represented by a line with a positive slope.

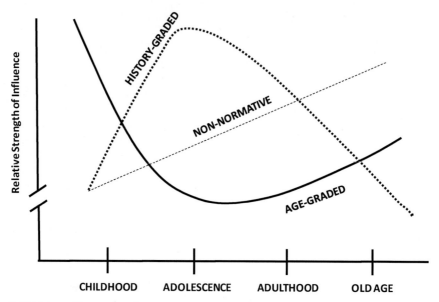

FIGURE 5-1 Illustrative influences of lifespan development.
NOTE: The interaction among the systems of influence is shown as linear and additive, but a transactional representation may turn out to be more useful.
SOURCE: Baltes, Reese, and Lipsitt, 1980; Figure 3. (Reprinted with permission.)

Lifespan development theories consider the impact of historical events on human development but also stress the importance of biological or cultural factors (e.g., social class, urban/rural status) in explaining the differences among people. This allowance for variance in the effects of historical events thus weakens the idea of generational membership based on normative historical events. Because the lifespan development perspective encompasses a broad range of factors that may influence a person's development, it is not a testable theory per se, but rather a paradigmatic framework (Baltes, Reese, and Lipsitt, 1980; Rudolph and Zacher, 2017). Moreover, although there has been much research on the effects of non-normative life events and individual differences on development, job performance, and motivation (Kanfer and Ackerman, 2004; National Academies of Sciences, Engineering, and Medicine, 2018; Schmidt and Hunter, 1998), no research has endeavored to examine all three of the above trajectories in lifespan development theories simultaneously.

Changes in Work Context

As discussed in Chapter 2, the world of work and the composition of the workforce are different from what they were in years past. These changes in the context of work, together with changes in the social context, have likely influenced the average person's attitudes, personality, values, and behaviors regardless of that person's age or generation. Thus, it is possible that as observers have witnessed changes in people's attitudes and behaviors, they have misattributed these changes to generations.

As noted in Chapter 4, when sophisticated methods are used to study differences among people, the observed differences are found to be related more strongly to period effects, indicating gradual change over time in the general population, than to generation effects, which would indicate change limited to a subgroup of the population based on birth year. Thus, the focus on generational issues misattributes actual changes in workplace context to changes in the preferences of workers from different generations. For instance, some have argued that younger generations prefer team-based work (Deal and Levinson, 2016); however, studies on this topic have showed that the personal preference for team-based work has actually decreased over time (Twenge et al., 2010). On the other hand, there is substantial evidence that work has become more interdependent and that teams are more prevalent in today's organizations (Wegman et al., 2018). Accordingly, future research is needed to identify and examine changes in contextual demands that impact the average worker (regardless of age or generation), instead of focusing predominantly on identifying differences in the work-related values of different generations of workers.

The Aging Workforce

Focusing on generational issues also has masked real challenges in the management of a more age-diverse workforce. New workplace norms, practices, and behaviors likely are developing as a function of changes in workforce demographics.[1] There is a large body of literature on "older" workers, and research attention to the aging workforce has increased as the percentage of workers over age 55 has continued to grow (Baltes, Rudolph, and Zacher, 2018; Fraccaroli and Truxillo, 2011; Hedge and Borman, 2012; Truxillo, Cadiz, and Hammer, 2015). This research has tended to focus on cognitive and physical changes as workers age and discrimination toward older workers (Hedge and Borman, 2012; Parry and McCarthy,

---

[1] Another National Academies study is reviewing the literature on the aging workforce in the United States. The study report, due in Spring 2021, will examine factors associated with decisions to continue working at older ages and the social and structural factors, including workplace policies and conditions, that inhibit or enable employment among older workers.

2017; Wang, Olson, and Shultz, 2013). Little research has focused on age diversity in the workplace compared with diversity issues around gender, race, and ethnicity, but there is growing interest in studying age as a dimension of diversity in the workplace and on work teams (Finkelstein et al., 2015; Truxillo, Cadiz, and Hammer, 2015).

In research to date on "older" workers, chronological age has been used as a primary measure in attempting to predict such organizational outcomes as individual performance, workplace discrimination, and benefits of age diversity. However, there is little agreement in the literature on what defines an older worker, and this literature has found considerable variation among older workers with regard to such attributes as cognitive functioning, future orientation, and personality (Bal et al., 2010; Mühlig-Versen, Bowen, and Staudinger, 2012; Nelson and Dannefer, 1992). This variation reflects the fact that the experience of aging is different for different people, and that aspects of work linked to older workers (e.g., retirement) vary across work contexts and cultures (Truxillo, Cadiz, and Hammer, 2015). Moreover, the research has yielded a number of contradictory or null findings. For example, if job performance does not decline with age, why does age discrimination persist? and Why is age diversity not linked consistently to productive outcomes? (North, 2019).

Research has shown that chronological age alone is not a sufficient predictor of organizational outcomes (Ng and Feldman, 2008). Potential paths forward for future research could entail considering the multidimensionality of age: to examine other work-related attributes in conjunction with chronological age, such as one's organizational tenure (Ng and Feldman, 2010a) and accumulated job experience and to consider the "age culture" of one's workplace. It may even be possible to consider the impact of generational perceptions in the workplace in conjunction with these other factors; that is, when workers consider themselves to be part of a generation, this perception may influence their work identity and attitudes (North, 2019).

SUMMARY

The goal of this chapter has been to explain that, while appealing, generational thinking has its risks and limitations when used to inform workforce management and employment practices. However, the persistence of generational stereotypes and biases can potentially create tensions in the workplace and impact employee decisions.

Research assessing the effects of generational biases in the workplace and advancing theories on their influences on workforce behavior continues to be worthwhile. Lifespan development perspectives might present a more adaptive path forward for exploring variation in individuals' work-related attitudes and behaviors than reliance on generational paradigms. Individual workers are constantly interacting with their environments, and these interactions can alter their work-related attitudes and behaviors in important ways.

A growing body of research is focused on managing workers with different needs and capabilities (e.g., considering flexibility in training opportunities). Another important line of inquiry is aimed at understanding the management approaches of effective organizations. Some of this work is discussed in Chapter 6. Overall, recent research has underscored the importance of work context and of the variations in needs among workers regardless of age or generation. Future research designed to inform workforce management needs to take an integrative approach, recognizing the importance of work context, shared influences, and individual trajectories.

> **Recommendation 5-1:** Researchers interested in examining relationships between work-related values and attitudes and subsequent behaviors and interactions in the workplace should endeavor to identify and better understand alternative explanations for observed outcomes that supplement explanations associated with generations. This effort would include attention to generational stereotypes and biases that might exist among workers. Research should also seek to better understand the multiple factors that influence attributes of individual workers, including aging in the workplace, and the changes in the work context that affect the behaviors of all workers.

6

Workforce Management
in a New Era

The purpose of this chapter is to respond to the last task of the committee's charge: to "provide conclusions and recommendations in terms of...changes that are warranted to better recruit and retain the best employees" (see Box 1-1 in Chapter 1). The committee was asked to review the research literature on generational attitudes and behaviors, and as discussed earlier in the report, this body of work is quite limited in terms of distilling evidence-based advice for workforce management. However, the committee's work in undertaking this study uncovered a number of challenges faced by employers and growing interest in improving workforce management. In addition, there are lessons to be learned about the process for developing and implementing effective workforce policies. This chapter provides alternative ways of thinking about workforce challenges perceived to be generational issues, and highlights the need for repeatable processes that allow employers to identify and address their workforce challenges.

As discussed in Chapter 2, the composition of the workforce is changing, leading to greater diversity among workers with regard to gender, race/ethnicity, and age. As a result, many employers today face the challenge of effectively maintaining and managing a diverse workforce, including employees of a wide range of ages, even as the nature of work is also changing. Faced with sometimes rapid changes in the job market, discontinuities in the rates of new hires and retirements, and evolving technologies that require new skills and communication approaches, employers are seeking ways to capitalize on the knowledge and experiences of a more diverse and aging workforce while responding to a new set of employee expectations around the conditions of work, including job flexibility, work–life balance, and professional development. Some general policies and practices appear to be effective in this regard across most organizations. For example, virtually all workers in the United States expect a safe working environment. Because there are many different types of employers and employees, however, each employer will also face its own unique set of workforce management challenges. In many cases, the effectiveness of specific practices will depend on such factors as the characteristics and size of the workforce; the culture of the organization; and the demands on and expectations for workers, as well as workers' own needs and expectations (Lawler and Boudreau, 2012).

The need for caution in applying generalizations with respect to workforce management is nowhere more evident than in the subject of this study. As discussed previously, empirical support is lacking for meaningful generational differences in work-related attitudes and behaviors. The most sophisticated research to date examining changes in work values over time in the United States attributes these changes more to evolving work contexts and aging processes than to differences among generations (e.g., Kalleberg and Marsden, 2019).

For employers contemplating revision of management practices to make them more effective, then, a generational perspective can be misguided, and may simply perpetuate stereotypes that likely do not apply to today's diverse workforces. Nor does a generational perspective reflect the individual differences in life and career experiences or in the abilities, attitudes, and values of the members of generational groups. For example, some 50-year-old workers will want to retire at age 55, while others will wish to work into their 70s. Policies and practices based on age, generation, or other personal characteristic are likely to be applied to some who do not desire or need a specific offering and exclude others who do.

> **Conclusion 6-1:** The notion of generational differences will continue to be appealing in the absence of compelling alternative explanations for real or perceived differences among people in the workplace. However, many of the stereotypes about generations result from imprecise use of the terminology in the popular literature and recent research, and thus cannot adequately inform workforce management decisions. Further, categorizing a group of workers by observable attributes can lead to overgeneralizations and improper assumptions about those workers, perhaps even discrimination.

This chapter builds on the discussion in Chapter 2, examining the implications for workforce management of the broad and discrete changes affecting the nature of work. The focus here is on recruitment and retention and the associated activities and policies that could help employers maintain qualified and motivated employees. First, the discussion draws on examples from selected employment sectors to illustrate workforce management challenges. It then highlights some of the workforce management opportunities afforded by the changing contexts of work. Next, the chapter considers the legal constraints on workforce management and how generation-based decisions could be interpreted in light of existing employment laws. The chapter concludes with the committee's recommendations for effective workforce management.

EXAMPLES OF WORKFORCE MANAGEMENT CHALLENGES

In undertaking this study, the committee sought to appreciate existing workforce challenges, particularly those that might be driving the need to understand generational differences. In so doing, we heard from personnel and readiness staff from the military, human resources consultants, and corporate speakers, and we reviewed numerous public documents on workforce issues. We recognized that all employers face some personnel issues that are unique to them and their industry and different from those of other workplaces, while some issues are similar across all workplaces. For example, scheduling challenges may be specific to employers with jobs that require a 24/7 physical presence in the workplace, but almost all employers must learn to manage an increasingly diverse workforce and successfully incorporate new technologies into their jobs and employment practices.

In general, improvements in U.S. labor market conditions (e.g., lower unemployment rates) coincide with greater difficulty in recruiting as well as increased turnover (Petrosky-Nadeau and Valletta, 2019), whereas a recession can have the opposite effect. In the period 2010 to 2019, the number of job openings nationwide increased by 138 percent (Work Institute, 2019). The unemployment rate was at a historically low level—3.6 percent in January 2020 (U.S. Bureau of Labor Statistics [BLS], 2020c); yet by April 2020, the unemployment rate had increased precipitously to 19.7 percent before decreasing to 13.3 percent in May (Rosenberg and Long, 2020).[1] Because the committee conducted much of its work during favorable job conditions, we heard a great deal about employers' concerns regarding recruitment and retention. Certainly, the unpredictability of employment during the COVID-19 pandemic and its longer-term impacts have created new workforce concerns for employers. Nevertheless, lessons can be learned from earlier responses to different challenges faced by employers.

---

[1] News media, such as the *Washington Post*, also reported that the Bureau of Labor Statistics made a "misclassification error" that likely underestimated the unemployment rate by 3 percentage points. Regardless of the actual numbers, the unemployment rate has increased dramatically in a short period of time (https://www.washingtonpost.com/business/2020/06/05/may-2020-jobs-report-misclassification-error).

The subsections that follow describe workforce challenges in greater detail for certain large employment sectors: recruitment in the military, retention in the military, resiliency for first responders, flexibility and delayed retirements in nursing, diversifying the workforce in education, and turnover in hospitality. These sectors were chosen in part because of the study sponsor (the U.S. Army Research Institute for the Behavioral and Social Sciences) and in part because of their prevalence in the generational literature. The examples described here provide a way to appreciate how broad shifts in the nature of work and the workforce have played out in different employment contexts.

Two points about these examples are worth noting. First, while the committee found a large amount of public information in the way of advice and training on managing different generations (see Chapter 3) and of research on generational differences aimed at informing employment practices (see Appendix A), we found no public documents linking specific human resources policies and practices to generational issues. Second, what works in one industry may not work as well in another or in the military context. For example, the military lacks some of the flexibility that private industry has to incentivize its recruits and personnel. But on the other hand, as a large employer, it has greater resources with which to research and evaluate solutions to the workforce problems it faces. All organizations, large and small, will have to consider the costs and benefits of investing in workforce strategies and determine how much effort should be expended.

Shifting Strategies for Military Recruitment

In recent years, the military services have been revisiting their marketing and recruiting strategies to address challenges with bringing qualified youth into their ranks (Grisales, 2019). For example, the Army missed its recruitment goal in 2018 for the first time since 2005, falling short by 6,500 soldiers (Cox, 2019). The focus of military recruiting for many years has been on marketing to youth and young adults to encourage people aged 17–24 to consider service in the military. Headlines and quotes from recruiting command personnel draw attention to efforts to appeal to members of "generation Z"; however, adjustments to recruitment strategies have less to do with different work values of young adults than with new, preferred avenues of communication. The military is targeting growing numbers of youth who lack family members with prior military service able to share information about military jobs, and is trying to capitalize on new communication platforms (Fadel and Morris, 2019; Myers, 2018). Leveraging new tools and forms of social interaction, the military services are using a variety of communication technologies to reach their intended audience, such as social media platforms (e.g., Instagram), videogames, and online advertising (Pawlyk, 2019). The Army also launched a new advertising campaign titled "What's Your Warrior?" to showcase various occupational opportunities it has to offer. In moving away from a combat-centric message, the Army aims to improve awareness of these opportunities and correct misconceptions about its jobs among today's youth and young adults (Brading, 2019).

The military must recruit hundreds of thousands of individuals each year, most of them young adults. All members of the military must meet specific requirements regarding health, education, moral character, and aptitude (U.S. Government Accountability Office [GAO], 2005). Nationwide, however, 71 percent of young people aged 17–24 do not qualify for military service. One factor contributing to the reduction in eligible recruits is the rising obesity rates among the U.S. population discussed in Chapter 2 (Maxey, Bishop-Josef, and Goodman, 2018).

Although broad societal trends and the realities of military life may reduce interest in military service and make it more difficult for the services to meet their recruiting goals (Wenger et al., 2019), the services can (and do) adjust their strategies and policies to address recruiting challenges. In general, the military services can, within some limits, change how stringently they set entry standards for new recruits (including which and how many conditions they waive at entry), how many recruiting resources (e.g., recruiters) they deploy, and how many recruits they are willing to enlist but hold until basic training slots become available. Congress also plays a role in trying to improve military recruitment and retention—for example, by enacting laws that authorize the services to provide educational benefits, retention bonuses, and the like.

Military Leave Options to Balance Work and Life Needs

In addition to recruitment, the military, like other organizations, wants to retain personnel who perform well and fill critical functions. In recent years, the military services have been implementing new approaches to talent management to respond to what have been termed "shifting generational values," as well as to remain competitive with industry (U.S. Army, 2019, p. 3). Promotion and monetary incentives have in the past been primary levers for retention. In recent years, however, the military also has developed a number of leave options, including family leave and sabbaticals, to allow its personnel to manage personal responsibilities and thereby reduce attrition.

Military service places significant demands on military personnel, including extended periods of time away from family (e.g., Segal, 1986). A recent National Academies report points to "a long history of evidence that families are important for military retention"; the report cites as an example the effect of spouses' support (or the lack thereof) for continued military service on service members' retention intentions (National Academies of Sciences, Engineering, and Medicine, 2019, p. 37). However, not all threats to retention affect everyone in the military workforce to the same degree. In the case of family influence on military service, the same National Academies report states that "family-related issues may be even more important to the retention of female than male service members," ostensibly because family obligations generally fall more on women than on men (2019, p. 38). Recent studies focused on the U.S. Air Force (Keller et al., 2018) and U.S. Coast Guard (Hall et al., 2019) provide further evidence of the challenges for female service members in balancing their military careers with family obligations, especially those related to childrearing.

The military and Congress have recently made changes to increase the amount of leave for those with primary caregiver responsibilities for newborn children, including children who are adopted. In 2016, then Secretary of Defense Ash Carter announced plans to extend paid maternity leave from 6 to 12 weeks and parental leave from 10 to 14 days (Ferdinando, 2016). Legislative changes followed: In the fiscal year 2017 National Defense Authorization Act (NDAA), Congress included authorization for the military to offer up to 12 weeks of leave for women who give birth, up to 6 weeks for individuals who are the primary caregiver following adoption, and up to 21 days for a secondary caregiver following either childbirth or adoption.[2]

The military services also have established programs that allow service members to take sabbaticals, or extended leaves of absence from active duty service, to pursue personal or professional goals (e.g., attain a college degree) or to meet personal-life needs (e.g., raise a family). Reasons cited for providing sabbaticals include a desire to retain talented service members amidst perceived competition from the private sector and perceived changes in societal values regarding work–life balance (see, e.g., Thie, Harrell, and Thibault, 2003).[3]

---

[2] See Section 521 of Public Law 114-328. These authorizations apply only to active duty personnel, personnel from the reserve component performing active Reserve or Guard duty, or reserve component personnel who may be recalled to active duty or mobilized for a period of 12 months or longer.

[3] Thie, Harrell, and Thibault (2003, p. 1) were asked by the Department of Defense (DoD) to examine the "feasibility and advisability of extended leaves for officers" in the military. The authors cite DoD's 2002 "Social Compact" as a "people-oriented motivation" for offering sabbaticals to military officers (p. 2). They note that this compact is "DoD's public acknowledgment that the department relies on a volunteer military in a changing context" that includes increasing numbers of young adults attending college and workers placing greater value on work–life balance (p. 2). They also indicate that DoD was concerned about private-sector competition for talent and wanted to provide workforce policies more in line with those of private-sector companies.

Although other sabbatical programs in the military may exist, the following are two that have been available for several years and are still offered. The first is the U.S. Coast Guard's Temporary Separation Program, or TEMPSEP, which was formally established in 2001 (Janaro, 2016).[4] TEMPSEP allows Coast Guard members on active duty to take up to 2 years of leave with the opportunity to return to active duty without an additional service obligation (Janaro, 2016). The second is the Career Intermission Pilot Program (CIPP), which was originally developed by the U.S. Navy. In the fiscal year 2009 NDAA, Congress authorized use of CIPP by all of the services.[5] This program allows a limited number of eligible active duty service members to take 3-year sabbaticals with the expectation of an additional service obligation upon return.[6] Members who participate in CIPP must affiliate with the Individual Ready Reserves, which allows for the provision of compensation and training during the sabbatical and facilitates the transition back to active duty (GAO, 2015).

Although TEMPSEP and CIPP were established to help the military services retain talent, service members have not utilized these programs as widely as expected. Recent studies have uncovered possible reasons for this underutilization: legal and military service restrictions limiting eligibility, and perceptions among service members that using the programs will negatively impact their military careers (e.g., less desirable assignments upon return and reduced promotion opportunity) (GAO, 2015; Hall et al., 2019). The U.S. Government Accountability Office (GAO) recommended that the Department of Defense (DoD) implement an evaluation plan to determine whether CIPP was meeting the Navy's retention goals for the program (GAO, 2015). While it is unclear whether DoD formally implemented an evaluation plan for CIPP, Congress has since removed its pilot program status and made it a permanent authority. It is now referred to by law as the Career Intermission Program (CIP).[7]

---

[4] According to Janaro (2016), TEMPSEP has its origins in the Care of Newborn Children (CNC) program established by the U.S. Coast Guard in 1991. The CNC program was aimed at retaining "high performing" women serving on active duty who had given birth.

[5] Congress authorized up to 20 officers and up to 20 enlisted members from each military service to enter CIPP each year. Congress also required that participants "have completed their initial active duty service agreement and are not currently receiving a critical skills retention bonus" and specified that for each month on sabbatical, the member owes 2 months of service upon return (GAO, 2015, p. 4). Congress extended the pilot program through 2019 in the fiscal year 2015 NDAA.

[6] Not only are eligibility requirements set in law, but each military service can impose additional eligibility requirements.

[7] See Section 551 of the fiscal year 2019 NDAA: https://www.congress.gov/115/crpt/hrpt676/CRPT-115hrpt676.pdf.

Resiliency for First Responders

The first responder industry comprises many occupations, including law enforcement, fire safety, emergency medical services, and other areas of public safety. As in many industries, first responder organizations are concerned about changes in the capability and experience of the workforce as older personnel leave or retire and are replaced by younger workers. This concern is driven in part by the fact that first responder jobs are often physical in nature, and first responders therefore tend to be young (Britton, 2018). Thus, the field must continually recruit and retain employees who can meet the physical demands of the work (Hanifen, 2017). The industry also faces challenges around integrating new technologies into its operations and building a diverse workforce (Federal Emergency Management Agency [FEMA], 2019). At the same time, moreover, there is growing recognition that changes in the work and threat environments may be increasing occupational stress, and organizations are seeking strategies for enhancing the well-being and resilience of the first responder workforce (Institute of Medicine, 2013, 2014). During the COVID-19 crisis, the pressures on first responders, particularly those who provide medical services, have exacerbated the stressful nature of their jobs (Cha, 2020; Wang, 2020).

In the popular literature, these workforce concerns are often expressed in terms of generations, as illustrated by government and industry articles titled "The Challenges of Managing Millennial Firefighters after Baby Boomers Retire" (Hanifen, 2017), "Generational Perspectives in Emergency Management" (Kirkland and Walsh, 2017), and "Millennials Might Just Be What the Fire Service Needs" (FEMA, 2019). Some of the advice offered on recruiting and retaining younger workers suggests adapting recruitment and training practices to bring them more in line with the stereotypes of millennials, including adjusting to such preferences and habits as use of social media and video and an emphasis on teamwork (Britton, 2018; FEMA, 2019; Wylie, 2017). Some advice focuses on minimizing the supposed worst traits of millennials (Eldridge, 2012) and some on leveraging the perceived strengths of younger workers. For example, an article published by the Federal Emergency Management Agency (FEMA) states that the traits of millennials—diverse, fitness-focused, and adept at technology—are exactly what fire departments need right now (FEMA, 2019), though as discussed in this report, the traits attributed to millennials are more likely representative of period effects affecting all workers more generally.

Flexible Scheduling and Delayed Retirements in Nursing

Shortages in the nursing profession continue to be a problem as the need for registered nurses intensifies in response to population aging and health care reforms (American Association of Colleges of Nursing, 2019; BLS, 2017). Although there have been some reports of furloughs of nurses during the COVID-19 pandemic, the demand for nurses remains greater than the supply (Dreher, 2020; Pitts, 2020). The accelerating rate of nurse retirements adds to the problem, posing risks for patient care, institutional memory, and leadership (Buerhaus et al., 2017; Stimpfel and Dickson, 2019). Health care organizations are considering strategies for retaining experienced nurses longer, attracting workers to the profession, and ensuring continuity in health care as nurses enter and leave the workforce. A number of articles in the generational literature focus on nursing (see Appendix A). While these studies fail to evaluate generational differences appropriately (see the discussion of methodological limitations in Chapter 4), it is interesting to reflect on what motivates such studies—the perception that understanding these differences is key to addressing issues of turnover, retention, management, leadership, job satisfaction, and occupational stressors.

Continuity in the nursing workforce and a balance in the numbers of experienced nurses and incoming professionals are important factors in enabling organizations to maintain the expertise required to provide specialized clinical and care management services. The transition of a newly licensed nurse from novice to expert takes several years (Benner, Tanner, and Chesla, 2013), and potentially longer for the shift from general nursing to a specialized field such as obstetrics or perioperative care. Employers of nurses thus face the dual challenges of retaining nurses nearing retirement age and rapidly developing and retaining the skills of their newly licensed counterparts.

Research has found that while their profession holds great intrinsic value for many nurses (e.g., satisfaction in providing care to patients), other factors, including pay, staffing levels, and support from nursing leadership, contribute to their retention across all age groups (Dols, Chargular, and Martinez, 2019; Stimpfel et al., 2019). Barriers to retaining nurses are also common across all generations (ages). They include (1) concerns about physical health, particularly during epidemics and notably among older nurses, but not limited to this age group because nurses of all ages can be physically limited from time to time; (2) inflexible schedules and long work hours (e.g., 12-hour shifts); (3) limited opportunities for retraining as health care technologies and tools change; and (4) concerns about occupational safety (Stimpfel et al., 2019; Wieck, Dols, and Landrum, 2010). Flexibility in schedules and/or duty assignments is one potential strategy for accommodating and retaining nurses with physical health issues (e.g., recovering from injury or surgery but able to work in some capacity) or nurses who need time to manage personal demands (e.g., care of children or elderly parents). Part-time positions or the option of split or reduced shifts of 4–6 hours could be an ideal situation for some nurses and manageable for some employers.

Education: Building a Diverse Teacher Workforce

Many states are reporting shortages of qualified teachers (Cross, 2017), and some scholars have argued that the educator workforce is affected by generational issues (e.g., Anthony, 2018; Lovely and Buffum, 2007; Petty, 2013). The difficulty of retaining relatively new teachers has at times been attributed to the notion that millennials' beliefs about work differ from those of previous generations (e.g., Anthony, 2018). However, few studies have examined teachers' work attitudes, and the one study the committee identified that compares work ethics among teachers uses a cross-sectional research design. Thus this study is not methodologically rigorous enough to support conclusions about an entire generation (see the discussion in Chapter 4). Other evidence points to the primacy of contextual factors (e.g., working conditions) over group differences.

Educators new to the profession today are more likely than those of the past to change careers and leave the profession within a few years. The predominant reason reported for leaving is being unhappy with work conditions (e.g., salary, student behavior, school leadership). However, turnover rates are not equivalent across schools, with about half of all teacher turnover in American public schools occurring in only 25 percent of schools. Higher turnover rates are associated with schools that are located in high-poverty neighborhoods and in urban or rural (as opposed to suburban) regions, and in those that have large minority populations (Ingersoll et al., 2018; Walker, 2019).

The makeup of the educator workforce has changed over time. First, it is aging, according to data collected by the National Center for Education Statistics since 1987. Even with increased retirements since 1987, the proportion of educators aged 50 and older rose from 20 percent in the 1987–1988 school year to 31 percent in the 2011–2012 school year (Ingersoll and Merrill, 2017). Moreover, the gender gap in the K–12 teaching workforce increased, with the number of female teachers greatly outnumbering that of male teachers (Ingersoll et al., 2018).[8]

In addition, although more teachers of diverse racial and ethnic backgrounds have entered the teaching workforce in recent years, there remains a large gap between the proportion of minority students in U.S. schools and the proportion of minority teachers in the educator workforce (Ingersoll et al., 2018). This gap is due primarily to a decrease in the number of white students that has coincided with an increase in the number of minority students, as opposed to a failure to recruit more diverse teachers. Specifically, the growth in minority teachers in the United States was three times greater

---

[8] Data describing the characteristics of the early childhood education workforce are limited and somewhat problematic compared with data on K–12 teachers (Workgroup on the Early Childhood Workforce and Professional Development, 2016).

than the growth in white teachers between 1987 and 2016. Increased recruitment and retention of teachers of color is an important need because research shows that when students of color learn from teachers of color, they accrue academic benefits; moreover, when the diversity of teachers in a school is higher, those teachers feel less isolated and are less likely to leave the school or the profession (Carver-Thomas, 2018).

Turnover in Hospitality

There are three primary workforce trends that present challenges for the hospitality industry: (1) the aging population places higher demands on leisure services from retirees and limits the pool of recruits to the industry, because it is dominated by younger workers (Baum, 2010); (2) its image of fun and glamor is positive for customers but can be unappealing to some as a career choice (Barron, Leask, and Fyall, 2014); and (3) the presence of mixed ages in the hospitality workforce has created tensions as older workers manage turnover among younger workers and aim to provide the best service to customers (Barron, Leask, and Fyall, 2014). A shortage of qualified staff in the hospitality industry can mean longer waits and fewer amenities for guests. The industry is challenged with very high turnover rates due to variations in staffing needs by season and a high proportion of young adults and students in the workforce (National Restaurant Association, 2019). The president of the American Hotel and Lodging Association has said that "the single greatest challenge facing our industry is filling essential jobs to ensure the quality service and amenities our guests expect and deserve" (Schwartz, 2019).

The response to the COVID-19 pandemic necessitated abrupt changes to the hospitality industry as many businesses were required to shut down or reduce services for a period of time. The unemployment rate in this industry skyrocketed to almost 40 percent in April 2020.[9] As conditions improve and businesses are allowed to offer a wider variety of personal services, the public's demand for leisure activities is likely to be high, and the hospitality industry will have to ramp back up quickly. It may also face new demands regarding occupational and public safety. Employers in the industry may once again turn to understanding the next generation of young workers.

Attention to generational groups in the hospitality industry has been offered as a potential way to understand, recruit, and retain employees (American Hotel and Lodging Association, 2016). Industry websites post such articles as "Attracting and Hiring Millennials for Your Restaurant"

---

[9] See workforce statistics at https://www.bls.gov/iag/tgs/iag70.htm.

and "4 HR Strategies for Optimizing a Multigenerational Workforce in the Hospitality Industry."[10]

A number of articles in the generational literature have focused on the hospitality industry (see Appendix A). These studies have attempted to further understand turnover (e.g., Brown, Thomas, and Bosselman, 2015), differences in work values (e.g., Chen and Choi, 2008), and job satisfaction (e.g., Kim, Knutson, and Choi, 2016). A systematic review of the generational research on hospitality workers makes observations similar to those in this report: the studies reviewed rely predominantly on cross-sectional research designs, and a paradigm shift is needed in how research considers generations of workers. The review finds that, much as in generational research in other contexts, age and period effects are minimally discussed, and little attention is paid to variances within and across generations that may better explain the complexity of generational issues in the hospitality industry (Sakdiyakorn and Wattanacharoensil, 2018).

WORKFORCE MANAGEMENT OPPORTUNITIES IN THE CHANGING WORLD OF WORK

The above examples illustrate some of the management challenges frequently mentioned in the literature reviewed for this study. The committee observed that many questions were raised about how to (1) manage the youngest generation of workers, which is touted as the historically most diverse group of workers; (2) recruit the new generation, as well as workers of varied ages; (3) train and develop workers from multiple generations (or more appropriately, those at different career stages); and (4) manage the varying needs of workers with regard to job flexibility. This section examines these management opportunities and some of the recent research with regard to capitalizing on a diverse workforce and increasing attention to recruitment strategies, professional development, and flexible work arrangements.

Diverse Workforce

Although it is difficult to directly attribute specific organizational benefits, including productivity and profitability, to a diverse workforce alone, having a diverse workforce offers obvious advantages, many of which likely do lead to productivity and profitability. Employers are more

---

[10] See https://www.gourmetmarketing.net/attracting-hiring-millennials-restaurant and https://www.trinet.com/insights/4-hr-strategies-for-optimizing-a-multigenerational-workforce-in-the-hospitality-industry.

likely to meet staffing needs when they recruit from the entire population of qualified applicants and not just a subset. Tapping into a larger pool of qualified workers from different racial, ethnic, gender, and age groups often provides access to talent with multiple perspectives, leading to better ideas and solutions and improved quality of business outputs under the right conditions (Ely and Thomas, 2001; Kamarck, 2019; Richard and Miller, 2013).

Many organizations strive to maintain a workforce that represents their customer base, finding that meeting societal expectations enhances the organization's reputation and is profitable (Holger, 2019; Hunt, Layton, and Prince, 2015; Lorenzo and Reeves, 2018; Zhang, 2020). In addition, complying with legal mandates has obvious benefits, especially if the organization is a federal contractor. Thus, the goal for many organizations is to form a workforce that represents a variety of experiences, cultures, and personal attributes. Proactive efforts to promote diversity and inclusivity, together with a strong commitment to diversity among senior management (Marquis et al., 2008), are important in creating a workforce that works together effectively to achieve organizational goals. Such efforts can in turn be self-reinforcing, helping to attract a more diverse workforce as well as retain existing talent (Buttner and Lowe, 2017; Li et al., 2020). Boehm, Kunze, and Bruch (2014), for example, found that a culture supportive of age diversity and fair implementation of employee benefits was a key factor in improving organizational performance and reducing reported intentions to quit.

While useful in many respects, a diverse workforce can also pose management challenges. Increased diversity often leads to perceptions of unfairness; research indicates, for example, that age diversity can create a climate of favoritism with negative effects (Kunze, Boehm, and Brunch, 2011). Diversity can also lead to tension in work teams and concerns about navigating training and management among the various groups of workers. Proactive attention to inclusive practices can help address such issues (Turban, Wu, and Zhang, 2019). Such practices are designed to enable all individuals and groups to feel welcome, valued, respected, and supported as they contribute to an organization's mission and success (see Box 6-1).

The best advice and research evidence recognize that there is no universal approach to increasing diversity and inclusion with respect to age or any other personal characteristic. Moreover, organizations have unique cultures that require specific strategies for their particular context. Organizations will benefit the most by developing a plan for achieving a diverse workforce and promoting inclusion that works for their own culture.

The heterogeneity of groups defined by demographic variables merits emphasizing. Whether defining groups by race, ethnicity, gender, or age,

BOX 6-1
Inclusion Practices

Inclusion practices are those that help workers feel

- psychologically and physically safe and comfortable in sharing different points of view;
- involved in teamwork and critical communications;
- respected and valued;
- influential in decision making;
- supported in sharing their authentic identity; and
- confident that the organization and management promote fair treatment, diversity, and employee growth.

SOURCE: Adapted from Shore, Cleveland, and Sanchez (2018).

there is no reason to believe that all members of a group share the same work values and needs. Consequently, a diverse workforce highlights the need for an organization to attend to an array of benefits, training, and worker accommodations that are adaptable for a number of individual circumstances in an organization's work environments.

Recruitment Strategies

The goal of personnel recruitment is to identify candidates whose preferences, skills, and abilities match the needs of the organization and the specific requirements of the job to be filled. The recruitment process is often separated into two phases: sourcing and attraction. Sourcing involves identifying candidates who are likely to possess the requisite skills for the job and to be interested in the job and the organization, while attraction refers to the process of developing and maintaining a candidate's interest throughout the hiring process.

The challenge of recruiting in labor markets characterized by low unemployment (as was the case prior to the COVID-19 pandemic) is finding sources for candidates and developing a process for attraction that will appeal to the broadest range of people who are likely to be viable candidates. When competition for people with specific skill sets is keen, organizations may need to pursue novel sources of candidates (Joyce, 2010). For example, companies that have long relied on campus recruiting from vocational technical schools, colleges, and universities may need to augment that approach by seeking to identify experienced candidates through such avenues as professional associations, job boards, employee referrals, internet search tools or job postings, and social media (Corporate Leadership Council and Recruiting Roundtable, 2006). It is also important to recognize that younger and older candidates may not be accessible from the same sources. As noted above, the military, which recruits primarily adolescents and young adults, has adapted to new communication media as they have emerged and gained popularity with these target candidates, whereas, other employers may have to leverage a variety of communication channels to reach workers of all ages. Moreover, the effectiveness of using any source of candidates depends at least in part on the nature of the job that needs to be filled. It is important to note as well that viable candidates will learn about opportunities from different sources and that many will use multiple sources. For example, a candidate may see an advertisement on television about a firm's goods or services and then go the organization's website to seek more information regarding career opportunities.

The benefits, culture, and environment offered by an organization, as well as its public image (Gatewood, Gowan, and Lautenschlager, 1993; Turban and Greening, 1997), will play a role in candidates' job decisions. In the course of this study, the committee learned from multiple articles and presenters (e.g., Deal, 2019) that the following features of work appear to be important to all workers, regardless of age or generation: appreciation, trust, career opportunities, clear goals and expectations, and fairness. Nonetheless, there are always differences in expectations of work among workers. What appeals to young adults and recent college graduates and to experienced workers, for example, may differ. Further, members of these two groups may also differ in multiple ways. Such factors as race/ethnicity and gender, as well as other demographic characteristics, geographic location, education level and college major, and work experiences, can be significant in determining what attracts different people to specific jobs.

Instead of making general assumptions about employee desires, then, some organizations are taking a more scientific approach to identifying what attracts individuals to an employer. Employee value proposition (EVP) is "the portfolio of tangible and intangible rewards an organization provides to employees in exchange for their job performance" (Shepherd, 2014, p. 580). EVP analysis, much of which is conducted at the organizational level, is designed to categorize employees into clusters based on similarity of desires. As part of this process, employees rank order a list of organizational characteristics and benefits according to importance. Then, those responses are used to cluster employees into groups with similar needs and desires. Importantly, this analysis can segment an organization's employee population according to the similarity of their perceptions of what is of value in the organization instead of according to age or generation. This information can then be used to communicate aspects of the organization that are likely to be rewarding to potential applicants. For example, a company might find in its EVP analyses that its education benefits are highly regarded by a cluster of employees that includes both new hires who are beginning their careers and desire more education to support them in their career development and more experienced employees who are looking to change roles and need new skills to be effective in their new positions. The organization might share the details of its education benefits and provide examples of how they have been used by different employees to achieve their career goals.

Professional Development

In light of the changing nature of work and the rapid development and implementation of new technologies (see Chapter 2), professional development has become more important, and training and retooling have become necessary at every career stage (Autor, 2015; Muro et al. 2017; Yoong and Huff, 2006). In addition, the number of nontraditional career paths is growing. Continuous learning environments support the development of new skills and the acquisition of new knowledge at any career stage and are critical to the success of organizations, as well as workers' development of professional skills, interests, and career identity (Hall and Mirvis, 1995, 2013).

Attention to professional development for employees can benefit both individual workers and the organization (Goldstein and Ford, 2002; Tannenbaum et al., 2010). Individuals develop knowledge and skills to perform their current jobs better, to advance their careers, and to prepare for new opportunities. The organization not only gains from employee development and the new and enhanced skills it provides to advance business objectives, but also can leverage professional development programs to ensure the transfer of institutional knowledge as its workforce evolves. In addition, development opportunities are often desirable to workers and therefore facilitate recruitment and retention.

Accompanying the increasing demand for professional development are changes in the nature of professional development activities within organizations. Organizations are increasingly shifting the responsibility for development from the employer to the employee. The employer provides resources and some guidance, but it is up to the employee to decide whether to take advantage of those opportunities. In addition, many organizations are shifting some of their training from formal, classroom settings to less formal, online, self-directed formats. Employees can no longer count on their organization to provide all of the skills and training they need to remain competitive for jobs in the 21st century. Rather, they must take it upon themselves to ensure that they have needed skills, and accordingly must seek out appropriate training and development opportunities (Hall and Mirvis, 1995, 2013).

A considerable body of research documents the importance of a conducive organizational climate with respect to formal training—one that signals to employees that learning is a valued activity (Armstrong-Stassen and Ursel, 2009). There is also growing recognition that the most effective career development experiences occur on the job, with reinforcement of the lessons learned through coaching from managers, mentors, or others (Tracey, Tannenbaum, and Kavanaugh, 1995). At the same time, each worker's expectations, interests, and decisions will be influenced by contextual factors both within and outside of work (Ackerman, 2000; Beier, Torres, and Gilberto, 2017). In this regard, there exist a number of stereotypes about generational groups and preferences for learning. For example, baby boomers, who are currently classified as the older generation of workers, are often characterized as lacking the desire and ability to learn continually at work (Posthuma and Campion, 2009). This perception may be the result of research findings regarding motivational and cognitive factors that do differentiate older and younger adult learners on average (Kubeck et al., 1996). However, cognitive abilities vary widely within age groups, and no assumptions about the importance of professional development to workers are warranted on the basis of age or generation alone (Hertzog et al., 2008).

Flexible Arrangements

The popular press paints a picture in which one generation values flexible schedules more than another. Depending on the source, these groups may be new hires and potential recruits or veteran employees. In reality, the evidence shows greater need for and acceptance of flexible schedules across ages and career stages.

Advances in technology and changes in the nature of job tasks have expanded the options for flexibility in work schedules, and research suggests that organizations are employing a range of strategies for creating flexible schedules (Chartered Institute of Personnel and Development, 2014). These options include job sharing, part-time work, temporary work, telecommuting, and flex scheduling (allowing employees to set their own hours for a given period). These options will not be appropriate for every employer, every job, or every employee. However, the response to the COVID-19 pandemic suggests that many employees and employers can quickly adapt to telecommuting and flexible work schedules. A careful analysis of such factors as job duties, equipment needed, people with whom interaction is required, and the nature of these interactions can help in determining whether flexible work schedules and locations are viable for particular employees and circumstances. In addition, some organizations provide guidelines for acceptable behavior when employees are working at different hours or remotely.

Adding flexibility to job structures offers many advantages, including increased employee well-being and job satisfaction (Lambert, Haley-Lock, and Henly, 2012; Moen et al., 2016a), reduced levels of turnover intentions (Moen et al., 2016b, 2017), and opportunities to delay retirements among more experienced employees and retain their critical skills/knowledge longer (Cahill, James, and Pitt-Catsouphes, 2015). Multiple surveys across different work sectors, as well as research examining work-related values (e.g., Twenge et al., 2010), show increasing interest in better work–life balance, and flexible schedules and work locations can enable individuals to balance work and life demands according to their needs.

DISCRIMINATION AND THE LAW

In the preceding sections, there have been several references to the misuse of a generational perspective for informing workforce management. When describing workforce challenges, the literature has often used generations as a proxy for different age groups. This section summarizes the legal constraints on workforce management and explains how generation-based decisions could be interpreted as age discrimination in light of existing employment laws.

Federal Protections

Federal laws currently in place prohibit making employment decisions on the basis of characteristics that include sex, race, color, national origin, religion, disability, or genetic information. Employers may not pay different wages to men and women who perform equal work, and they may not discriminate against a person because of pregnancy and childbirth. In addition, people over age 40 may not be discriminated against based on age. These federal laws, which include Title VII of the Civil Rights Act, Title I of the Americans with Disabilities Act, the Pregnancy Discrimination Act, the Equal Pay Act, and the Age Discrimination in Employment Act (ADEA), are enforced by the Equal Employment Opportunity Commission (EEOC). Each of the laws operates slightly differently, based on the text of the legislation as well as how it has been interpreted by courts and the EEOC. In general, employment decisions—including recruitment, hiring, firing, and promotion—may not be made based on characteristics that define a protected class of people, and employers may not retaliate against a person who complained about such discrimination or filed or participated in a lawsuit about the discrimination. In addition, the laws prohibit employment practices that apply to all but have a disproportionately negative effect on people of a certain race, color, religion, sex, national origin, or disability unless the practice is job-related and necessary to the operation of the business. An employment practice that applies to all but has a disproportionately negative effect on people aged 40 or older is prohibited if the practice is not based on a reasonable factor other than age.[11]

The Age Discrimination in Employment Act

When age, generational categories, or stereotypes about generations are used in the workplace to make decisions or policies, the employer may potentially be in violation of the ADEA, as well as various state laws. The ADEA was enacted in 1967 as the result of congressional concern that older workers were facing discrimination based on age, including arbitrary age limits for certain jobs (29 U.S.C. § 621). Congress considered including age in Title VII of the Civil Rights Act of 1964, the landmark employment law that prohibits discrimination on the basis of race, color, religion, sex, and national origin, but opted to create a separate law instead. Over the years, the ADEA and Title VII have developed slightly different requirements and legal standards, with age discrimination cases in general being more difficult to prove than Title VII cases (Dunleavey et al., 2019).

The ADEA applies to employers with 20 or more employees and to employees or applicants who are aged 40 and older (29 U.S.C. § 631). (In practice, evidence suggests that age discrimination occurs much later than age 40. Many of the ADEA cases highlighted by the EEOC involve employees in their 50s and 60s.[12]) The ADEA does not apply to uniformed personnel in the military services. The law prohibits discrimination based on age in any area of employment, including hiring, promotion, pay, assignments, layoffs, training, benefits, and firing (29 U.S.C. § 623). There are several ways in which employers could potentially violate the ADEA with age- or generation-based decisions, each discussed below.

Disparate Treatment

Making employment-related decisions about ADEA-covered applicants or employees based solely on age violates the letter of the ADEA. Examples of these types of decisions include choosing to hire or promote a younger rather than an older worker because of the age of the employees involved and not their job-relevant knowledge, skills, abilities, and other characteristics, or choosing to lay off older workers before younger ones solely on the

---

[11] See https://www.eeoc.gov/age-discrimination.
[12] See https://www.eeoc.gov/eeoc/litigation/selected/adea.cfm.

basis of age. The Supreme Court has clarified that age, rather than some other factor, must actually motivate the employer's decision. For example, a decision to fire an older employee before his pension benefits vest does not violate the ADEA if the decision was based on years of service rather than the employee's actual age (*Hazen Paper Co. v. Biggins*, 507 U.S. 604 [1993]). The Court noted, however, that if the claimed factor were merely a proxy for age, the employer could potentially be liable (Id.).

By most accounting, the oldest members of the millennial generation are nearing age 40, the age at which protection under ADEA begins. However, the Supreme Court has held that workers—even if they are 40 or older—are protected under the ADEA only if a discriminatory action favors younger workers at the expense of older workers (*General Dynamics Land Systems Inc. v. Cline*, 540 U.S. 581 [2004]). For example, if an employer promoted a 55-year-old employee rather than a 45-year-old employee based on age, the 45-year-old employee would not have a claim under the ADEA. However, if the employer promoted a 35-year-old rather than the 45-year-old based on age, the 45-year-old would have a cognizable claim. The Supreme Court noted that Congress intended to protect older workers from unfair preferences benefiting younger workers, and said that if Congress had intended the act to protect younger workers, it would not have excluded everyone under 40. The takeaway from the *General Dynamics* case is that employers may, if they choose, always favor older over younger workers without violating the ADEA.

In practice, disparate treatment cases can be difficult to prove. While older workers might feel that they were discriminated against because of age, their employer might claim that the decision was made because a younger worker had certain skills or abilities. For example, the younger worker might have been perceived to be more innovative than the older worker. If a court found that the employment decision was based on this factor rather than age alone, the discrimination claim would not succeed.

Disparate Impact

Policies that apply to everyone but have a disparate negative impact on older workers can violate the ADEA (*Smith v. City of Jackson, Miss.*, 544 U.S. 228 [2005]). For example, if an employer required employees to pass a screening test that few or no older employees could pass (e.g., a physical ability test evaluating strength), this action could violate the ADEA unless the test was based on a reasonable factor other than age (e.g., job requirements for strength). Unlike a disparate treatment violation, a disparate impact violation is based on the effect on the employee (or applicant), not on the employer's motivations (Id.) To combat a disparate impact claim, an employer must demonstrate that the policy was based on a "reasonable factor other than age" (*Smith v. City of Jackson, Miss.*, 544 U.S. 228 [2005]). The burden is on the employer to show reasonableness (*Meacham v. Knolls Atomic Power Laboratory*, 554 U.S. 84 [2008]). In response to *Smith* and *Meacham*, the EEOC has clarified that for an employment practice to be based on a "reasonable factor other than age," it must be "reasonably designed and administered to achieve a legitimate business purpose in light of the circumstances, including its potential harm to older workers."[13]

It should be noted that a workplace benefit that is designed to benefit younger employees but is available to all employees likely does not violate the ADEA, as long as there is a reasonable factor other than age behind the policy (Dunleavey et al., 2019). For example, a company might be able to demonstrate that a policy or practice (e.g., child care, videogame room) is common among its competitors, and thus it needs to offer the same benefit to stay competitive even though child care and videogames may appeal more to younger than older employees.

New technologies being used for hiring and management—for example, online assessments, targeted online recruitment, and the use of artificial intelligence for assessment—could potentially be the source of ADEA violations. While these technologies are too new to have been thoroughly tested under the ADEA, there are signs that this is an area in which employers could be vulnerable. In 2019, the EEOC determined that seven companies had violated the ADEA by blocking older workers from viewing Facebook job advertisements. As part of a settlement with civil rights organizations, Facebook has agreed to change the ways in which advertisers can target users (Gillum and Tobin, 2019). The use of artificial intelligence for making employment decisions may also be an area ripe for ADEA violations; the use of data about previously successful candidates or employees is likely only to perpetuate existing biases against older workers. For example, if previous hiring decisions favored younger workers, this bias would be baked into the data used to build an artificial intelligence system (e.g., "successful employees are under age 35"), and the new system would reflect the age bias present in the previous hiring practices.

There is disagreement as to whether disparate impact claims can be brought by applicants as well as employees under the ADEA. Other provisions of the ADEA explicitly refer to "applicants," whereas the plain language of the disparate impact provision in the law refers only to "employees"; it prohibits an employer from "limit[ing], segregat[ing], or classify[ing] his employees in any way which would deprive or tend to deprive any individual of employment opportunities or otherwise adversely affect his status as an employee, because of such individual's age" (29

---

[13] See https://www.eeoc.gov/laws/regulations/adea_rfoa_qa_final_rule.cfm#_ftn.

U.S.C. § 623[a][2]). The 7th U.S. Circuit Court of Appeals (*Kleber v. CareFusion Corporation*, No. 17-1206 [7th Cir. Jan. 23, 2019]) and the 11th U.S. Circuit Court of Appeals (*Villarreal v. R.J. Reynolds Tobacco Co.*, 839 F.3d 958 [2016]) have both ruled that disparate impact claims may not be brought by applicants, because of the plain language of the statute. However, the U.S. District Court for the Northern District of California has allowed a class action suit from a group of applicants to move forward (*Rabin v. PricewaterhouseCoopers LLP*, 236 F.Supp.3d 1126 [N.D. Cal. 2017]), and the District Court for the Southern District of Texas also refused to dismiss an applicant's disparate impact claim (*Champlin v. Manpower Inc.*, No. 4:16-CV-00421 [S.D. Tex. Jan. 24, 2018]). As a result of this circuit split, disparate impact claims are currently available to applicants in some regions of the country but not in others; further guidance from Congress, other federal courts, or the Supreme Court would be required to resolve this conflict.

Stereotypes

An employer may violate the ADEA if an employment-related decision is based on stereotypes, generalizations, or stigmas associated with workers of a certain age. In fact, as the Supreme Court noted, this concern was the driving force behind the ADEA; Congress recognized that age discrimination was based on unsupported stereotypes, and that empirical evidence demonstrated that the performance of older workers was at least as good as that of younger workers (*EEOC v. Wyoming*, 460 U.S. 226 [1983]). The ADEA requires employers to make decisions based on an employee's "merits, and not their age" (*Western Air Lines, Inc. v. Criswell*, 472 U.S. 400, 422 [1985]), and employers cannot use age as a proxy for other characteristics, such as productivity (*Hazen Paper Co. v. Biggins*, 507 U.S. 604 [1993]). The Court noted that a decision based on a reasonable factor other than age is one that does not rely on stereotypes or generalizations, even if the reasonable factor is related to age (e.g., years of service) (*Hazen Paper Co. v. Biggins*, 507 U.S. 604 [1993]).

Harassment

The text of the ADEA does not explicitly prohibit harassment of employees based on age. However, several circuit courts have held that employees can sue employers for creating a hostile work environment based on age (*Milan v. Dediol v. Best Chevrolet, Inc., et al.*, No. 10-30767 [5th Cir. 2011]; *Crawford v. Medina General Hosp.*, 96 F.3d 830 [6th Cir. 1996]). Hostile work environment claims have long been available under Title VII of the Civil Rights Act, which prohibits discrimination based on race, color, religion, sex, or national origin (although again, harassment is not explicitly prohibited in the text of the law) (*Meritor Savings Bank, FSB v. Vinson*, 477 US 57 [1986]). The 5th Circuit set out a four-part test for a violation of the ADEA due to age-based harassment in the workplace:

- the employee is over age 40;
- the employee was subjected to harassment, either through words or actions, based on age;
- the nature of the harassment was such that it created an objectively intimidating, hostile, or offensive work environment; and
- there exists some basis for liability on the part of the employer.

The EEOC has clarified its interpretation of when age-based harassment violates the ADEA, stating that harassment can include offensive or derogatory remarks about a person's age, and that harassment is illegal when "it is so frequent or severe that it creates a hostile or offensive work environment or when it results in an adverse employment decision."[14] Like harassment under Title VII, the hostile work environment need not be created directly by the employer; harassment by supervisors, coworkers, and even customers may lead to a violation of the ADEA.

State Laws

While the ADEA is rather narrow—applying only to employees over 40, and only to discrimination against older in favor of younger workers—a number of state laws take a broader view of age discrimination. For example, Oregon Revised Statute 659A.030 prohibits employers from using age as the basis for hiring or firing decisions unless the decision is based on "a bona fide occupational qualification reasonably necessary to the normal operation of the employer's business." The law applies to any individual aged 18 or older and does not require a showing that the discriminatory practice favors younger at the expense of older workers. Most states have laws that prohibit discrimination in employment based on age, although many of these laws mirror the ADEA by restricting claims to those aged 40 and over.[15]

Application to Generations

The ADEA and state laws prohibit discrimination based on age and do not explicitly address generational categories or stereotypes. However, a court could find that an employer that had made a decision based on

---

[14] See https://www.eeoc.gov/laws/types/age.cfm.

[15] See https://www.ncsl.org/research/labor-and-employment/discrimination-employment.aspx.

an employee's generation was using generation as a mere proxy for age, placing the employer in violation of federal or state law. For example, an employer's refusal to hire a baby boomer based solely on this generational category would almost certainly violate age discrimination laws. Employment decisions based on stereotypes about generations—such as refusing to put workers of a certain generation in a specific job position—could be particularly vulnerable to ADEA claims because the congressional intent behind the ADEA was to combat these types of pervasive stereotypes and stigmatization of older workers. Workplace harassment based on age or generational stereotypes could also be fertile ground for an ADEA claim; for example, a hostile work environment created through frequent comments, jokes, and insults about workers in a particular generation could violate the ADEA. While federal disparate treatment and disparate impact claims require that the discrimination be against an older in favor of a younger worker, there is no such requirement for a harassment claim. It is feasible, though not yet tested, that millennials reaching the age of 40 could make claims of workplace harassment based on comments, jokes, and insults about their generation (e.g., that millennials are entitled or lazy). In addition, workers of younger generations may have recourse under state laws if they are discriminated against based on their generational category.

Hiring Practices and Selection Tests

Federal and state laws have largely made the selection and management practices of employers less biased against a group defined by demographic characteristics. For example, it is no longer permissible to deny a promotion to someone based on his or her race, to refuse to hire a woman because she is pregnant, or to fire an employee once he or she reaches a certain age.

Selection practices, dictated in part by these laws, have benefited and improved from many years of research and its incorporation into practice. Developments in the evaluation of potential job performance have now made it possible for organizations to consider performance broadly; identify the aspects of performance of greatest relevance; reduce the likelihood that selection decisions are biased on applicants' age, gender, race, or ethnicity, among other characteristics; and adjust selection systems accordingly (Campbell et al., 1993; Rotundo and Sackett, 2002). The goal of effective selection practices that are lawful is to identify candidates who are likely to succeed because they possess the required knowledge, skills, abilities, and other characteristics. For virtually all jobs, race/ethnicity, gender, and age are not characteristics that define capability. Good selection tools identify capable candidates regardless of their demographic characteristics.

RECOMMENDATIONS FOR EFFECTIVE
WORKFORCE MANAGEMENT

To develop effective management strategies, employers must first establish policies and procedures that are fair to all employees and serve the organization's goals. Once such practices are in place, management must then continuously examine the changing context of work in the workplace and employees' needs, and develop policies that continue to meet those needs and accord well with the employer's mission and job requirements. (See the summary of this chapter's key messages in Box 6-2). Organizations can best evaluate their workforce management decisions if they have structures and procedures in place for regularly collecting and analyzing workforce data. Employers should be able to determine what their employees find attractive about the organization, what reasons are behind turnover, and what their employees need and want to be satisfied and successful in their jobs and careers. An array of options is most likely to best serve a diverse workforce and give employees and their managers the flexibility to decide what work arrangements, training, and benefits are of most value to individual workers and jobs.

Recommendation 6-1: In considering approaches to workforce management, employers and managers should focus on the needs of individual workers and the changing contexts of work in relation to job

---

BOX 6-2
The Changing World of Work and Implications
for Employment Practices: Key Messages

The nature of work and the workforce is changing. Each employer will face its own unique set of challenges as a result of these broad changes.

An organization's workforce management policies and practices and the needs of employees should be revisited regularly.

Collecting and maintaining workforce data will be important to determine what policies and practices are likely to be effective and to assess their effectiveness.

Changes and developments in recruitment strategies and management policies and practices should align with an organization's values, mission, and goals and the needs of its employees.

Tailoring management policies and employment practices to a specific group defined by some personal characteristic, such as generation, is unlikely to meet the needs of all members of that group and may exclude others unfairly. In general, management policies, such as those related to work arrangements and benefits, should apply to all workers, allowing each to decide whether to take advantage of them.

requirements instead of relying on generational stereotypes. Employers can be guided in making any needed changes to employment practices and policies by a thorough assessment of changes in their own work environment, job requirements, and human capital.

The committee's review of the generational literature and supplemental study of the changing nature of work uncovered several workforce challenges that we believe broadly affect most workplaces and organizations in the United States. These include recent challenges with the recruitment and retention of workers; increasing diversity in the labor force; rising demand for better work–life balance across all ages of workers; and the need to reexamine training and professional development in light of changing employer–employee relationships, the incorporation of new technologies, and greater proportions of team-based approaches to work.

The task for an organization is not to find a single, permanent answer to the workforce challenges listed above. The nature of these challenges changes over time. As a result of the economic effects of the COVID-19 pandemic, for example, the recruiting challenges of January 2020 were substantially different from those 4 months later in May. Moreover, employees' needs and values change with broader societal changes. For example, the need for elder care has increased over time because of a number of factors, including longevity in the United States; the mobility of the American worker; and women's entry into the workforce in large numbers, which makes them unavailable for caregiving. In addition, possible solutions are constantly evolving. For example, recently developed teleconferencing tools have enhanced the effectiveness of remote working and facilitated flexible work schedules and locations. Organizations must then evaluate the new policies and procedures they undertake to determine their impact on organizational effectiveness and the extent to which employees' needs are met. Thus, the committee recommends that organizations develop effective ways to regularly identify changes in the context of the work environment and employees' needs and values, determine currently available solutions to any challenges arising from these changes that are aligned with the organization's mission and values and meet the needs of employees, and evaluate those solutions. In other words, the solution to these workforce challenges is not a particular one-time solution but a repeatable process by which challenges are identified and resolved and the solutions to those challenges are evaluated.

Recommendation 6-2: Employers should have processes in place for considering and reevaluating on a regular basis an array of options for workforce management, such as policies for recruiting, training and development, diversity and inclusion, and retention. The best options will be consistent with the organization's mission, employees, customer base, and job requirements and will be flexible enough to adjust to different worker needs and work contexts as they change.

References

Acemoglu, D., and Autor, D.H. (2011). Skills, tasks, and technologies: Implications for employment and earnings. In O. Ashenfelter and D. Card (Eds.), *Handbook of Labor Economics* (Vol. 4, pp. 1043–1171). Elsevier.

Ackerman, P.L. (2000). Domain-specific knowledge as the "dark matter" of adult intelligence: Gf/Gc, personality and interest correlates. *Journals of Gerontology: Series B: Psychological Sciences and Social Sciences, 55*(2), P69–P84.

Allport, G.W. (1954/1979). *The Nature of Prejudice*. New York: Doubleday.

American Association of Colleges of Nursing. (2019). *Nursing Shortage*. Washington, DC. Available: https://www.aacnnursing.org/News-Information/Fact-Sheets/Nursing-Shortage.

American Hotel and Lodging Association. (2016). *Millennials in the Hotel Industry*. Available: https://www.ahla.com/sites/default/files/Millennial_Retention_Survey.pdf.

American Psychological Association (2017). *2017 Work and Well-Being Survey*. Available: https://www.apaexcellence.org/assets/general/2017-work-and-wellbeing-survey-results.pdf.

Anthony, A.P. (2018). "Missing" millennials and the great workforce divide: School leaders lament: Why are we unable to retain our young teachers? *School Administrator, 75*(11), 23–25. Available: http://my.aasa.org/AASA/Resources/SAMag/2018/Dec18/Anthony.aspx.

Appelbaum, E., and Batt., R. (2017). The networked organization: Implications for jobs and inequality. In D. Grimshaw, C. Fagan, G. Hebson, and I. Tavbors (Eds.), *Making Work More Equal: A New Labour Market Segmentation Approach*. Manchester, UK: Manchester University Press.

Armstrong-Stassen, M., and Ursel, N.D. (2009). Perceived organizational support, career satisfaction, and the retention of older workers. *Journal of Occupational and Organizational Psychology, 82*(1), 201–220. Available: https://doi.org/10.1348/096317908X288838.

Army Science Board. (2015). *Talent Management and the Next Training Revolution*. Washington, DC: Department of the Army. Available: https://apps.dtic.mil/dtic/tr/fulltext/u2/1063616.pdf.

Autor, D.H. (2015). Why are there still so many jobs? The history and future of workplace automation. *Journal of Economic Perspectives* 29(3), 3–30.

Autor, D.H. (2019a). *The Changing Geography of Work, Wages, and Skills: 1960–2015.* Presentation to the Committee on the Consideration of Generational Issues in Workforce Management and Employment Practices on July 30, 2019, Washington, DC. Available: https://www.nationalacademies.org/our-work/consideration-of-generational-issues-in-workforce-management-and-employment-practices.

Autor, D.H. (2019b). Work of the past, work of the future. *AEA Papers and Proceedings,* 109:1-32.

Bal, P.M., Jansen, P.G., Van Der Velde, M.E., de Lange, A.H., and Rousseau, D.M. (2010). The role of future time perspective in psychological contracts: A study among older workers. *Journal of Vocational Behavior,* 76(3), 474–486.

Baltes, P.B., Reese, H.W., and Lipsitt, L.P. (1980). Life-span developmental psychology. *Annual Review of Psychology* 31:65–110.

Baltes, B., Rudolph, C., and Zacher, H. (Eds.). (2018). *Work Across the Lifespan.* New York: Academic Press.

Barron, P., Leask, A., and Fyall, A. (2014). Engaging the multi-generational workforce in tourism and hospitality. *Tourism Review,* 69(4), 245–263.

Baum, T. (2010). Demographic changes and the labour market in the international tourism industry. In I. Yeoman, C. Hsu, K. Smith, and S. Watson (Eds), *Tourism and Demography* (pp. 1–19). Oxford, UK: Goodfellow.

BBC News. (2014, March 1). *The Original Generation X.* Available: https://www.bbc.com/news/magazine-26339959.

Beier, M.E., Torres, W.J., and Gilberto, J.M. (2017). Continuous development throughout a career: A lifespan perspective on autonomous learning. In J.E. Ellingson and R.A. Noe (Eds.), *Autonomous Learning in the Workplace: SIOP Organizational Frontier Series* (pp. 179–200). New York: Routledge.

Bell, A., and Jones, K. (2018). The hierarchical age–period–cohort model: Why does it find the results that it finds? *Quality and Quantity,* 52(2), 783–799.

Benner, P., Tanner, C.A., and Chesla, C.A. (2013). *Expertise in Nursing Practice: Caring, Clinical Judgment, and Ethics, 2nd edition.* New York: Springer Publishing Company.

Bennett, N., and G.J. Lemoine. (2014). What VUCA really means for you. *Harvard Business Review.* Available: https://hbr.org/2014/01/what-vuca-really-means-for-you.

Bernstein, J. (2006). *All Together Now: Common Sense for a Fair Economy.* San Francisco: Berrett-Koehler Publishers.

Blalock, H.M. (1967). Status inconsistency, social mobility, status integration. *American Sociological Review,* 32, 790–801.

BLS (U.S. Bureau of Labor Statistics). (2017). *Employment Projections – 2016-26.* Washington, DC: Department of Labor. Available: https://www.bls.gov/news.release/archives/ecopro_10242017.pdf.

———. (2019a). *Employment by Major Industry Sector.* Washington, DC: Department of Labor. https://www.bls.gov/emp/tables/employment-by-major-industry-sector.htm.

———. (2019b). *Labor Force Characteristics by Race and Ethnicity, 2018.* Washington, DC: Department of Labor. Available: https://www.bls.gov/opub/reports/race-and-ethnicity/2018/home.htm.

———. (2019c). *Time Use of Millennials and Nonmillennials.* Washington, DC: Department of Labor. Available: https://www.bls.gov/opub/mlr/2019/article/time-use-of-millennials-and-nonmillennials.htm.

———. (2020a). *All Employees, Manufacturing [MANEMP],* retrieved from FRED, Federal Reserve Bank of St. Louis. Available: https://fred.stlouisfed.org/series/MANEMP.

———. (2020b). *Economic News Release USDL-20-0108. Union members – 2019.* Available: https://www.bls.gov/news.release/pdf/union2.pdf.

———. (2020c). Unemployment rate 2.0 percent for college grads, 3.8 percent for high school grads in January 2020. *The Economics Daily.* Available: http://www.bls.gov/opub/ted/2020/unemployment-rate-2-percent-for-college-grads-3-8-percent-for-high-school-grads-in-january-2020.htm.

Boehm, S.A., Kunze, F., and Bruch, H. (2014). Spotlight on age-diversity climate: The impact of age-inclusive HR practices on firm-level outcomes. *Personnel Psychology,* 67(3), 667–704.

Brading, T. (2019, November 13). "What's your Warrior?"—Army's new recruiting effort targets gen Z. *Army News Service.* Available: https://www.army.mil/article/229645/whats_your_warrior_armys_new_recruiting_effort_targets_gen_z.

Bresman, H., and Rao, V.D. (2017, August 25). A survey of 19 countries shows how generations X, Y, and Z are—and aren't—different. *Harvard Business Review.* Available: https://hbr.org/2017/08/a-survey-of-19-countries-shows-how-generations-x-y-and-z-are-and-arent-different.

Brewer, M., and Feinstein, A. (1999). Dual processes in the cognitive representation of persons and social categories. In S. Chaiken and Y. Trope (Eds.), *Dual Processes in Social Pyschology* (pp. 255–270). New York: Guilford Press.

Brink, K.E., Zondag, M.M., and Crenshaw, J.L. (2015). Generation is a culture construct. *Industrial and Organizational Psychology: Perspectives on Science and Practice,* 8(3), 335–340.

Britton, S.J. (2018). Understanding and embracing the multigenerational workforce. *Journal of Emergency Medical Services.* Available: https://www.jems.com/2018/07/17/understanding-embracing-the-multigenerational-workforc.

Brooks, D. (2000, November 5). What's the matter with kids today? Not a thing. *The New York Times.* Available at: https://www.nytimes.com/2000/11/05/books/what-s-the-matter-with-kids-today-not-a-thing.html.

Brown, E.A., Thomas, N., and Bosselman, R.H. (2015). Are they leaving or staying: A qualitative analysis of turnover issues for Generation Y hospitality employees with a hospitality education. *International Journal of Hospitality Management,* 46, 130–137.

Brown, T.A. (2015). *Confirmatory Factor Analysis for Applied Research (2nd Edition).* New York: Guilford Publications.

Buerhaus, P.L., Skinner, L.E., Auerbach, D.L., and Staiger, D.O. (2017). Four challenges facing the nursing workforce in the United States. *Journal of Nursing Regulation,* 8(2), 40–46.

Burnett, J. (2011). *Generations: The Time Machine in Theory and Practice.* Surrey, Burlington: Ashgate Publishing Company.

Buttner, E.H., and Lowe, K. (2017). The relationship between perceived pay equity, productivity, and organizational commitment for US professionals of color. *Equality, Diversity and Inclusion: An International Journal,* 36, 73–89. doi: 10.1108/EDI-02-2016-0016.

Cahill, K.E., James, J.B., and Pitt-Catsouphes, M. (2015). The impact of a randomly assigned time and place management initiative on work and retirement expectations. *Work, Aging and Retirement,* 1, 350–368.

Campbell, J.P., McCloy, R.A., Oppler, S.H., and Sager, C.E. (1993). A theory of performance. In N. Schmitt and W.C. Borman and Associates (Eds.), *Personnel Selection in Organizations* (pp. 35–70). San Francisco, CA: Jossey-Bass Publishers.

Campbell, S.M., Twenge, J.M., Campbell, W.K. (2017). Fuzzy but useful constructs: Making sense of the differences between generations. *Work, Aging and Retirement,* 3(2), 130–139. Available: https://doi.org/10.1093/workar/wax001.

Campbell, W.K., Campbell, S.M., Siedor, L.E., and Twenge, J.M. (2015). Generational differences are real and useful. *Industrial and Organizational Psychology: Perspectives on Science and Practice*, 8(3), 324–331.

Cappelli, P. (1999). *The New Deal at Work: Managing the Market-Driven Workforce*. Boston, MA: Harvard Business School Press.

Cappelli, P.H., and Keller, J.R. (2013a). Classifying work in the new economy. *Academy of Management Review*, 38(4), 575–596.

———. (2013b). A study of the extent and potential causes of alternative employment arrangements. *Industrial and Labor Relations Review*, 66(4), 874–901.

Cappelli, P., Bassi, L., Katz, H., Knoke, D., Osterman, P., and Useem, M. (1997). *Change at Work*. New York: Oxford University Press.

Carver-Thomas, D. (2018). *Diversifying the Teaching Profession: How to Recruit and Retain Teachers of Color*. Palo Alto, CA: Learning Policy Institute.

Casey, L. (2016, October 31). Minding the generation gap: Investigating media portrayals of millennials and 'Gen Z'. *New York Times*. Available: https://www.nytimes.com/2016/10/28/learning/lesson-plans/minding-the-generation-gap-investigating-media-portrayals-of-millennials-and-gen-z.html.

Cha, A.E. (2020, April 20). When COVID-19 claimed two of their own, these EMTs grieved and kept on going. *The Washington Post*. Available: https://www.washingtonpost.com/health/when-covid-19-claimed-two-of-their-own-these-emts-grieved-and-carried-on/2020/04/20/200c9542-81c5-11ea-a3ee-13e1ae0a3571_story.html.

Champlin v. Manpower Inc., No. 4:16-CV-00421 (S.D. Tex. Jan. 24, 2018). Available: https://casetext.com/case/champlin-v-manpower-inc-1.

Chartered Institute of Personnel and Development. (2014). *HR: Getting Smart about Agile Working*. London: CIPD. Available: https://www.cipd.co.uk/Images/hr-getting-smart-agile-working_2014_tcm18-14105.pdf.

Chen, P.J., and Choi, Y. (2008). Generational differences in work values: A study of hospitality management. *International Journal of Contemporary Hospitality Management*, 20(6), 595–615.

Cheremukhin, A. (2014). Middle-skill jobs lost in U.S. Labor Market. *Economic Letter, Federal Reserve Bank of Dallas*, 9(5), 1–4. Available: https://www.dallasfed.org/~/media/documents/research/eclett/2014/el1405.pdf.

Cherlin, A.J. (2014). *Labor's Love Lost: The Rise and Fall of the Working-Class Family in America*. New York: Russell Sage Foundation.

Cillufo, A. (2019). 5 facts about student loans. *Factank: News in the Numbers*. Washington, DC: Pew Research Center. Available: https://www.pewresearch.org/fact-tank/2019/08/13/facts-about-student-loans.

Clark, B., Joubert, C., and Maurel, A. (2014). The career prospects of overeducated Americans. *Working Paper 20167, NBER Working Paper Series*. Cambridge, MA: National Bureau of Economic Research.

Cohn, R. (1972). On interpretation of cohort and period analyses: A mathematical note. In M.W. Riley, M. Johnson, and A. Foner (Eds.), *Aging and Society: A Sociology of Age Stratification, Vol. III*. New York: Russell Sage Foundation.

Computing Technology Industry Association. (2019). *Cyberstates 2019: The Definitive Guide to the U.S. Tech Industry and Tech Workforce*. Downers Grove, IL: CompTIA Properties, LLC. Available: https://www.cyberstates.org/pdf/CompTIA_Cyberstates_2019.pdf.

Comte, A. (1830–1840). *Cours de Philosophie Positive*.

Cooper, R.N. (2008). Global imbalances: Globalization, demography, and sustainability. *Journal of Economic Perspectives*, 22(3), 93–112.

Copland, D. (1991). *Generation X: Tales for an Accelerated Culture*. New York: St. Martins Press.

Corporate Leadership Council and Recruiting Roundtable. (2006). *Creating a Passive Candidate Recruiting Strategy.* Arlington, VA: CEB.

Costa, D.L., and M.E. Kahn. (1999). Power couples: Changes in the locational choice of the college educated, 1940–1990. *NBER Working Paper 7109.* Cambridge, MA: National Bureau of Economic Research. Available: https://www.nber.org/papers/w7109.

Costanza, D.P., and Finkelstein, L.M. (2015). Generationally based differences in the workplace: Is there a *there* there? *Industrial and Organizational Psychology 8*(3), 308–323. doi: https://doi.org/10.1017/iop.2015.15.

———. (2017). Generations, age, and the space between: Introduction to the special issue. *Work, Aging and Retirement, 3,* 109–112.

Costanza, D.P., Badger, J.M., Fraser, R.L., Severt, J.B., and Gade, P.A. (2012). Generational differences in work-related attitudes: A meta-analysis. *Journal of Business and Psychology,* 27(4), 375–394.

Costanza, D.P., Darrow, J.B., Yost, A.B., and Severt, J.B. (2017). A review of analytical methods used to study generational differences: Strengths and limitations. *Work, Aging, and Retirement 3*(2), 149–165. doi: https://doi.org/10.1093/workar/wax002.

Cox, M. (2019, March 21). *Army Scaling Back Recruiting Goals After Missing Target, Under Secretary Says.* Available: https://www.military.com/daily-news/2019/03/21/army-scaling-back-recruiting-goals-after-missing-target-under-secretary-says.html.

Crawford v. Medina General Hosp., 96 F.3d 830 (6th Cir. 1996).

Creswell, J.W. (2015). *A Concise Introduction to Mixed Methods Research.* Thousand Oaks, CA: Sage.

Cross, F. (2017). *Teacher Shortages Areas, Nationwide Listing 1990-1991 through 2017-2018.* Washington, DC: U.S. Department of Education. Available: https://www2.ed.gov/about/offices/list/ope/pol/bteachershortageareasreport201718.pdf.

Cunningham, D. (2014, August). Now hear this–Millennials bring a new mentality: Does it fit? *U.S. Naval Institute Proceedings,* 140/8/1,338. Available: https://www.usni.org/magazines/proceedings/2014/august/now-hear-millennials-bring-new-mentality-does-it-fit.

Darley, J.M., and Gross, P.H. (1983). A hypothesis-confirming bias in labeling effects. *Journal of Personality and Social Psychology,* 44(1), 20–33. doi: https://doi.org/10.1037/0022-3514.44.1.20.

Deal, J. (2019). *Generational Differences in the Workplace.* Presentation to the Committee on the Consideration of Generational Issues in Workforce Management and Employment Practices on May 29, 2019, Washington, DC. Available: https://www.nationalacademies.org/our-work/consideration-of-generational-issues-in-workforce-management-and-employment-practices.

Deal, J., and Levinson, A.R. (2016). *What Millennials Want from Work: How to Maximize Engagement in Today's Workforce.* New York: McGraw-Hill Education.

DeBickes, A., and Stiller, K. (2016). *Annotated Bibliography for Generational Differences, 2008-2015 Report No. 22-16.* Patrick AFB, FL: DEOMI Press. Available: https://www.deomi.org/DownloadableFiles/research/documents/Generational-Differences.pdf.

Desilver, D. (2019). 5 facts about the national debt. *Fact Tank: News in the Numbers.* Washington, DC: Pew Research Center. Available: https://www.pewresearch.org/fact-tank/2019/07/24/facts-about-the-national-debt.

Dey, M., and Loewenstein, M.A. (2020). How many workers are employed in sectors directly affected by COVID-19 shutdowns, where do they work, and how much do they earn? *Monthly Labor Review.* U.S. Bureau of Labor Statistics. doi: https://doi.org/10.21916/mlr.2020.6.

Dimock, M. (2019). Defining generations: Where millennials end and generation Z begins. *Fact Tank: News in the Numbers.* Pew Research Center. Available: https://www.pewresearch.org/fact-tank/2019/01/17/where-millennials-end-and-generation-z-begins.

Dols, J.D., Chargular, K.A., and Martinez, K.S. (2019). Cultural and generational considerations in RN retention. *Journal of Nursing Administration*, 49(4), 201–207.

Donnelly, K., Twenge, J., Clark, M., Shaikh, S., Beiler-May, A., and Carter, N. (2016). Attitudes toward women's work and family roles in the United States, 1976–2013. *Psychology of Women Quarterly*, 40(1), 41–54.

Dreher, A. (2020, May 6). From shortages to furloughs, pandemic gives in-demand nursing profession a stress test. *The Spokesman-Review*. Available: https://www.spokesman.com/stories/2020/may/06/from-shortages-to-furloughs-pandemic-gives-in-dema.

Dunleavey, E., Lustenberger, D., Hennen, M., and Willner, K. (2019). Legal Perspective. Presentation to the Committee on the Consideration of Generational Issues in Workforce Management and Employment Practices on October 3, 2019, Washington, DC. Available: https://www.nationalacademies.org/our-work/consideration-of-generational-issues-in-workforce-management-and-employment-practices.

Durham, W.H. (1991). *Coevolution: Genes, Culture and Human Diversity*. Stanford, CA: Stanford University Press.

Dwyer, R.E., and Wright, E.O. (2019). Low-wage job growth, polarization, and the limits and opportunities of the service economy. *Russell Sage Foundation Journal of the Social Sciences*, 5, 56–76.

EEOC v. Wyoming, 460 U.S. 226 (1983). Available: https://tile.loc.gov/storage-services/service/ll/usrep/usrep460/usrep460226/usrep460226.pdf.

Elder, G.H., Jr. (1974). *Children of the Great Depression: Social Change in Life Experience*. Chicago, IL: University of Chicago Press.

Elder, G.H., Jr. (1985). *Life Course Dynamics: Trajectories and Transitions, 1968-1980*. Ithaca, NY: Cornell University Press.

Elder, G.H., Jr. (1998). The life course as developmental theory. *Child Development*, 69, 1–12.

Elder, G.H., Jr., Kirkpatrick-Johnson, M., and Crosnoe, R. (2003). The emergence and development of life course theory. In J.T. Mortimer and M. J. Shanahan (Eds.), *Handbook of the Life Course* (pp. 3–19). New York: Springer.

Eldridge, L. (2012). *How to Work with the Next Generation of Police Leadership–The Millennials*. Available: https://www.policeone.com/police-jobs-and-careers/articles/how-to-work-with-the-next-generation-of-police-leadership-the-millennials-YtOrzYV-7wEiry1ME.

Elmendorf, D.W., and Sheiner, L.M. (2017). Federal budget policy with an aging population and persistently low interest rates. *Journal of Economic Perspectives*, 31(3), 175–194. doi: https://doi.org/10.1257/jep.31.3.175.

Ely, R.J., and Thomas, D.A. (2001). Cultural diversity at work: The effects of diversity perspectives on work group processes and outcomes. *Administrative Science Quarterly*, 46, 229–273.

Enam, A., and Konduri, K.C. (2018). Time allocation behavior of twentieth-century American generations: GI generation, silent generation, baby boomers, generation X, and millennials. *Transportation Research Record*, 2672(29), 69–80.

Fadel, L., and Morris, A. (2019). After falling short, U.S. Army gets creative with new recruiting strategy. *National Public Radio*. Available: https://www.npr.org/2019/01/06/682608011/after-falling-short-u-s-army-gets-creative-with-new-recruiting-strategy.

Fannon, Z., and Nielsen, B. (2019). Age-period-cohort models. *Economics and Finance*. doi: https://doi.org/10.1093/acrefore/9780190625979.013.495.

FEMA (Federal Emergency Management Agency). (2019). Millenials might just be what the fire service needs. *The InfoGram*, 19(26), 2. Available: https://www.usfa.fema.gov/operations/infograms/072519.html.

Ferdinando, L. (2016, January 28). Carter announces 12 weeks paid military maternity leave, other benefits. *U.S. Department of Defense News*. Available: https://www.defense.gov/Newsroom/News/Article/Article/645958/carter-announces-12-weeks-paid-military-maternity-leave-other-benefits.

Ferry, D. (2015). *The Odes of Horace*. New York: Farrar, Straus and Giroux.

Finkelstein, L.M., Ryan, K.M., and King, E.B. (2013). What do the young (old) people think of me? Content and accuracy of age-based metastereotypes. *European Journal of Work and Organizational Psychology*, 22(6), 633–657.

Finkelstein, L.M., Truxillo, D.M., Fraccaroli, and Kanfer, R. (2015). *Facing the Challenges of a Multi-Age Workforce: A Use-Inspired Approach*. New York: Routledge Press.

Finkelstein, L.M., Voyles, E.C., Thomas, C.L., and Zacher, H. (2020). A daily diary study of responses to age meta-stereotypes. *Work, Aging, and Retirement*, 6(1), 28–45.

Fischer, C.S., and Hout, M. (2006). *Century of Difference: How America Changed in the Last One Hundred Years*. New York: Russell Sage Foundation.

Fiske, S.T. (1998). Stereotyping, prejudice, and discrimination. In D.T. Gilbert, S.T. Fiske, and G. Lindzey (Eds.), *Handbook of Social Psychology* (Vol. 2, pp. 357–411). Boston, MA: McGraw-Hill.

Fiske, S.T., and Taylor, S.E. (1991). *Social Cognition, 2nd ed*. New York: McGraw-Hill.

Fosse, E., and Winship, C. (2019). Analyzing age-period-cohort data: A review and critique. *Annual Review of Sociology*, 45, 467–492.

Fraccaroli, F., and Truxillo, D.M. (2011). Special issue call for papers: Age in the workplace: Challenges and opportunities. *European Journal of Work and Organizational Psychology*, 20(5), 727–727.

Fried, Y., Levi, A.S., and Laurence, G. (2008). Job design in the new world of work. In S. Cartwright and C.L. Cooper (Eds.), *Oxford Handbook of Personnel Psychology*: 587–597. New York: Oxford Press.

Fry, R. (2018). Millennials are the largest generation in the U.S. labor force. *Fact Tank: News in the Numbers*. Pew Research Center. Available: https://www.pewresearch.org/fact-tank/2018/04/11/millennials-largest-generation-us-labor-force.

Fry, R., and Parker, K. (2018). Early benchmarks show post-millennials on track to be most diverse, best-educated generation yet. *Social and Demographic Trends*. Pew Research Center. Available: https://www.pewsocialtrends.org/2018/11/15/early-benchmarks-show-post-millennials-on-track-to-be-most-diverse-best-educated-generation-yet.

GAO (U.S. Government Accountability Office). (2005). *DOD Needs Action Plan to Address Enlisted Personnel Recruitment and Retention Challenges. GAO-06-134*. Washington, DC. Available: https://www.gao.gov/assets/250/248566.pdf.

———. (2015). *DOD Should Develop a Plan to Evaluate the Effectiveness of Its Career Intermission Pilot Program. GAO-16-35*. Washington, DC.

Gatewood, R.D., Gowan, M.A., and Lautenschlager, G.J. (1993). Corporate image, recruitment image, and initial job choice decisions. *The Academy of Management Journal*, 36(2), 414–427.

General Dynamics Land Systems Inc. v. Cline, 540 U.S. 581 (2004). https://tile.loc.gov/storage-services/service/ll/usrep/usrep540/usrep540581/usrep540581.pdf.

Gentile, B., Campbell, W.K., and Twenge, J.M. (2010). Birth cohort differences in self-esteem, 1988-2008: A cross-temporal meta-analysis. *Review of General Psychology*, 14(3), 261–268.

———. (2013). Generational cultures. In A.B. Cohen (Ed.), *Culture Reexamined: Broadening Our Understanding of Social and Evolutionary Influences* (pp. 31–48). Washington, DC: American Psychological Association.

Georges, C. (1993, September 12). The boring twenties. *Washington Post*. Available: https://www.washingtonpost.com/archive/opinions/1993/09/12/the-boring-twenties/001a040b-d7f2-46ee-b489-44b68b8f82bb.

Gergen, K.J. (2014). Pursuing excellence in qualitative inquiry. *Qualitative Psychology* 1, 49–60. doi: http://dx.doi.org/10.1037/qup0000002.

Gerras, S.J. (2010). *Strategic Leadership Primer, 3rd Edition*. Department of Command, Leadership, and Management. Carlisle Barracks, PA: Army War College. Available: https://publications.armywarcollege.edu/pubs/3516.pdf.

Gilleard, C. (2004). Cohorts and generations in the study of social change. *Social Theory and Health*, 2, 106–119. doi: https:/doi.org/10.1057/palgrave.sth.8700023.

Gillum, J., and Tobin, A. (2019, March 19). Facebook won't let employers, landlords or lenders discriminate in ads anymore. *ProPublica*. Available: https://www.propublica.org/article/facebook-ads-discrimination-settlement-housing-employment-credit.

Golden, L., Henly, J.R., and Lambert, S. (2013). Work schedule flexibility for workers: A path to employee happiness? *Journal of Social Research and Policy* 2(4), 107–134.

Goldstein, I.L., and Ford, K. (2002). *Training in Organizations: Needs Assessment, Development, and Evaluation (4th Edition)*. Belmont, CA: Wadsworth.

Graen, G., and Grace, M. (2015). New talent strategy: Attract, process, educate, empower, engage and retain the best. *SHRM-SIOP Science of HR White Paper Series*. Society for Human Resource Management and Society for Industrial and Organizational Psychology. Available: https://www.siop.org/Portals/84/docs/SIOP-SHRM%20White%20Papers/SHRM-SIOP_New_Talent_Strategy.pdf.

Greenhouse, S. (2019). *Beaten Down, Worked Up: The Past, Present and Future of American Labor*. New York: Penguin Random House.

Grisales, C. (2019). Military recruitment, retention challenges remain, service chiefs say. *Stars and Stripes*. Available: https://www.stripes.com/military-recruitment-retention-challenges-remain-service-chiefs-say-1.581364.

Hacker, J.S. (2006). *The Great Risk Shift: The Assault on American Jobs, Families, Health Care, and Retirement and How You Can Fight Back*. New York: Oxford University Press.

Hackman, J.R., and Oldham, G.R. (1975). Development of the job diagnostic survey. *Journal of Applied Psychology*, 60(2), 159–170.

Hall, G.S. (1904). *Adolescence: Its Psychology and Its Relations to Physiology, Anthropology, Sociology, Sex, Crime, Religion and Education*. New York: D. Appleton and Company.

Hall, D.T., and Mirvis, P.H. (1995). The new career contract: Developing the whole person at midlife and beyond. *Journal of Vocational Behavior*, 47(3), 269–289. doi: https:/doi.org/10.1006/jvbe.1995.0004.

———. (2013). Redefining work, work identity, and career success. In D.L. Blustein (Ed.), *The Oxford Handbook of the Psychology of Working* (pp. 203–217). New York: Oxford University Press.

Hall, K.C., Keller, K.M., Schulker, D., Weilant, S., Kidder, K.L., and Lim, N. (2019). *Improving Gender Diversity in the U.S. Coast Guard: Identifying Barriers to Female Retention*. Homeland Security Operational Analysis Center. Available: https://www.rand.org/content/dam/rand/pubs/research_reports/RR2700/RR2770/RAND_RR2770.pdf.

Hanifen, R. (2017, July 14). Challenges of managing millennial firefighters after baby boomers retire. *EDM Digest*. Available: https://edmdigest.com/opinion/firefighters-retire.

Harris, K.M. (2010). An integrative approach to health. *Demography*, 47(1), 1–22.

Harvey, D. (2005). *A Brief History of Neoliberalism*. Oxford, UK: Oxford University Press.

Haskel, J., Lawrence, R.Z., Leamer, E.E., and Slaughter, M.J. (2012). Globalization and U.S. wages: Modifying classic theory to explain recent facts. *Journal of Economic Perspective* 26(2), 119–140.

Hazen Paper Co. v. Biggins, 507 U.S. 604 (1993). https://tile.loc.gov/storage-services/service/ll/usrep/usrep507/usrep507604/usrep507604.pdf.

Hedge, J.W., and Borman, W.C. (2012). *The Oxford Handbook of Work and Aging.* New York: Oxford University Press.

Hertzog, C., Kramer, A.F., Wilson, R.S., and Lindenberger, U. (2008). Enrichment effects on adult cognitive development: Can the functional capacity of older adults be preserved and enhanced? *Psychological Science in the Public Interest, 9*, 1–65. doi: https://doi.org/10.1111/j.1539-6053.2009.01034.x.

Hipple, S.F. (2016). *Labor Force Participation: What Has Happened Since the Peak?* Washington, DC: U.S. Bureau of Labor Statistics. Available: https://www.bls.gov/opub/mlr/2016/article/labor-force-participation-what-has-happened-since-the-peak.htm.

Hobcraft, J., Menken, J., and Preston, S. (1985). Age, period, and cohort effects. In *Demography: A Review.* New York: Springer.

Hoffman, B. Shoss, M., and Wegman, L. (2020). The changing nature of work and workers (introduction). In Hoffman, Shoss, and Wegman (Eds.), *The Cambridge Handbook of the Changing Nature of Work and Workers* (pp. 3–19). Cambridge: Cambridge University Press.

Hogan, H., Perez, D., and Bell, W. (2008). *Who (Really) Are the First Baby Boomers.* Washington, DC: U.S. Census Bureau.

Hogan, P.I., Santomier, J., Jr., and Myers, B. (2016). Sport education in the VUCA world. *Journal of Physical Education and Sports Management, 3*(1), 1–37.

Hogler, R.L. (2020). The rise and decline of organized labor in the United States: American unions from Truman to Trump. In B. Hoffman, M. Shoss, and L. Wegman (Eds), *The Cambridge Handbook of the Changing Nature of Work and Workers* (pp. 173–191). Cambridge: Cambridge University Press.

Holger, D. (2019, October 26). The business case for more diversity. *The Wall Street Journal.* Available: https://www.wsj.com/articles/the-business-case-for-more-diversity-11572091200.

Horn, J.L., and McArdle, J.J. (1992). A practical and theoretical guide to measurement invariance in aging research. *Experimental Aging Research 18*(3), 117–144.

Howe, N., and Strauss, W. (1993). *13th Gen: Abort, Retry, Ignore, Fail?* New York: Vintage Books.

———. (2000). *Millennials Rising: The Next Great Generation.* New York: Vintage.

———. (2007). The next 20 years: How customer and workforce attitudes will evolve. *Harvard Business Review* (July–August), 41–52.

Howell, D.R., and Kalleberg, A.L. (2019). Declining job quality in the United States: Explanations and evidence. *Russell Sage Foundation Journal of the Social Sciences 5*(4), 1–53.

Hunt, V., Layton, D., and Prince, S. (2015). Why diversity matters. McKinsey & Company. Available: https://www.mckinsey.com/business-functions/organization/our-insights/why-diversity-matters#.

Hunt, J., and Nunn, R. (2019). Is employment polarization informative about wage inequality and is employment really polarizing? *IZA Discusson Papers,* No. 12472. Available: https://www.econstor.eu/bitstream/10419/202818/1/dp12472.pdf.

Inceoglu, I., Segers, J., and Bartram, D. (2012). Age related differences in work motivation. *Journal of Occupational and Organizational Psychology, 85*(2), 300–329.

Ingersoll, R., and Merrill, L. (2017). *A Quarter Century of Changes in the Elementary and Secondary Teaching Force: From 1987 to 2012. Statistical Analysis Report (NCES 2017-092).* U.S. Department of Education. Washington, DC: National Center for Education Statistics. Available: http://nces.ed.gov/pubsearch.

Ingersoll, R.M., Merrill, E., Stuckey, D., and Collins, G. (2018). Seven trends: The transformation of the teaching force–Updated October 2018. *Consortium for Policy Research in Education Research Reports*. Available: https://repository.upenn.edu/cpre_researchreports/108.

Institute of Medicine. (2013). *A Ready and Resilient Workforce for the Department of Homeland Security: Protecting America's Front Line*. Washington, DC: The National Academies Press. doi: https://doi.org/10.17226/18407.

———. (2014). *Advancing Workforce Health at the Department of Homeland Security: Protecting Those Who Protect Us*. Washington, DC: The National Academies Press. doi: https://doi.org/10.17226/18574.

Jacobs, T.O. (2002). *Strategic Leadership: The Competitive Edge*. Washington, DC: National Defense University, Industrial College of the Armed Force.

Janaro, S. (2016, June 24). Human capital strategy: Updates to the temporary separation program. *Coast Guard All Hands*. Available: https://allhands.coastguard.dodlive.mil/2016/06/24/human-capital-strategy-updates-to-the-temporary-separation-program.

Jenkins, R. (2018). *This Is Why Millennials Care So Much About Work-Life Balance*. Available: https://www.inc.com/ryan-jenkins/this-is-what-millennials-value-most-in-a-job-why.html.

———. (2019). *Leading the Four Generations at Work*. American Management Association. Available: https://www.amanet.org/articles/leading-the-four-generations-at-work.

Johns, G. (2006). The essential impact of context on organizational behavior. *Academy of Management Review* 31, 386–408.

Joshi, A., Decker, J.C., and Franz, G. (2011). Generations in organizations. *Research in Organizational Behavior*, 31, 177–205.

Joyce, L.W. (2010). Building the talent pipeline: Attracting and recruiting the best and brightest. In R. Silzer and B. Dowell (Eds.), *Strategy-Driven Talent Management: A Leadership Imperative*. San Francisco, CA: John Wiley & Sons, Inc.

Jürges, H. (2003). Age, cohort, and the slump in job satisfaction among West German workers. *Labour*, 17(4), 489–518.

Kalleberg, A.L. (2000). Nonstandard employment relations: Part-time, temporary, and contract work. *Annual Review of Sociology* 26, 341–365.

———. (2011). *Good Jobs, Bad Jobs: The Rise of Polarized and Precarious Employment Systems in the United States, 1970s-2000s*. New York: Russell Sage Foundation, American Sociological Association Rose Series in Sociology.

———. (2018). *Precarious Lives: Job Insecurity and Well-being in Rich Democracies*. Cambridge, UK: Polity Press.

Kalleberg, A.L., and Marsden, P.V. (2019). Work values in the United States: Age, period, and generational differences. *Annals of the American Academy*, 682(1), 43–59.

Kamarck, K.N. (2019). *Diversity, Inclusion, and Equal Opportunity in the Armed Services: Background and Issues for Congress*. CRS Report R44321. Washington, DC: Congressional Research Service.

Kanfer, R., and P.L. Ackerman. (2004). Aging, adult development, and work motivation. *The Academy of Management Review* 29(3), 440–458. doi: https://doi.org/10.2307/20159053.

Keister, L., Zucman, G., Shapiro, T., Pfeffer, F., Bartels, L., Sherraden, M., Kopczuk, W., and Scheve, K. (2015). *Rising Wealth Inequality: Causes, Consequences and Potential Responses*. Policy Brief. University of Michigan Poverty Solutions. Available: https://poverty.umich.edu/research-projects/policy-briefs/rising-wealth-inequality-causes-consequences-and-potential-responses.

Keller, K.M, Hall, K.C., Matthews, M., Payne, L.A., Saum-Manning, L., Yeung, D., Schulker, D., Zavisian, S., and Lim, N. (2018). *Addressing Barriers to Female Officer Retention in the Air Force.* Santa Monica, CA: RAND. Available: https://www.rand.org/pubs/research_reports/RR2073.html.

Keyes, K.M., and Li, G. (2012). Age–period–cohort modeling. In G. Li and S.P. Baker (Eds.), *Injury research: Theories, methods, and approaches* (pp. 409–426). New York: Springer.

Kim, M., Knutson, B.J., and Choi, L. (2016). The effects of employee voice and delight on job satisfaction and behaviors: Comparison between employee generations. *Journal of Hospitality Marketing and Management, 25*(5), 563–588.

Kirkland, H., and Walsh, W. (2017). *Generational Perspectives in Emergency Management.* Federal Emergency Management Agency. Available: https://training.fema.gov/hiedu/docs/generational%20perspectives%20final%205.22.17docx.pdf.

Kleber v. CareFusion Corporation, No. 17-1206 (7th Cir. Jan. 23, 2019). https://cases.justia.com/federal/appellate-courts/ca7/17-1206/17-1206-2019-01-23.pdf?ts=1548280830.

Koning, P., and Raterink, M. (2013). Re-employment rates of older unemployed workers: Decomposing the effect of birth cohorts and policy changes. *Economist (Netherlands) 161*(3), 331–348.

Kooij, D.T., De Lange, A.H., Jansen, P.G., Kanfer, R., and Dikkers, J.S. (2011). Age and work related motives: Results of a meta analysis. *Journal of Organizational Behavior, 32*(2), 197–225.

Kowske, B.J., Rasch, R., and Wiley, J. (2010). Millennials' (lack of) attitude problem: An empirical examination of generational effects on work attitudes. *Journal of Business and Psychology, 25*(2), 265–279.

Krippner, G. (2005). The financialization of the American economy. *SocioEconomic Review 3,* 173–208.

Kubeck, J.E., Delp, N.D., Haslett, T.K., and McDaniel, M.A. (1996). Does job-related training performance decline with age? *Psychology and Aging, 11*(1), 92–107.

Kunze, F., Boehm, S.A., and Bruch, H. (2011). Age diversity, age discrimination climate and performance consequences—a cross organizational study. *Journal of Organizational Behavior, 32,* 264-290.

Kupperschmidt, B.R. (2000). Multigeneration employees: Strategies for effective management. *The Manager, 19,* 65–76.

Lambert, S.J., Haley-Lock, A., and Henly, J.R. (2012). Schedule flexibility in hourly jobs: Unanticipated consequences and promising directions. *Community, Work and Family 15*(3), 293–315.

Lawler, E.E., and Boudreau, J.W. (2012). *Effective Human Resource Management: A Global Analysis.* Palo Alto, CA: Stanford University Press.

Lee, J.M., Pilli, S., Gebremariam, A., Keirns, C.C., Davis, M.M., Vijan, S., Freed, G.L. Herman, W.H., and Gurney, J.G. (2010). Getting heavier, younger: Trajectories of obesity over the life course. *International Journal of Obesity, 34*(4), 614–623.

Lee, H., Lee, D., Guo, G., and Harris, L.M. (2011). Trends in body mass index in adolescence and young adulthood in the United States: 1959-2002. *Journal of Adolescent Health, 49*(6), 601–608.

Lerner, R.M. (2002). *Concepts and Theories of Human Development, 3rd ed.* Mahwah, NJ: Erlbaum.

Lester, S.W., Standifer, R.L., Schultz, N.J., and Windsor, J.M. (2012). Actual versus perceived generational differences at work: An empirical examination. *Journal of Leadership and Organizational Studies, 19*(3), 341–354.

Leuty, M.E., and Hansen, J.I.C. (2014). Teasing apart the relations between age, birth cohort, and vocational interests. *Journal of Counseling Psychology, 61*(2), 289–298.

Levitt, H.M., Bamberg, M., Creswell, J.W., Frost, D.M., Josselson, R., and Suárez-Orozco, C. (2018). Journal article reporting standards for qualitative primary, qualitative meta-analytic, and mixed methods research in psychology: The APA Publications and Communications Board task force report. *American Psychologist* 73, 26–46.

Li, S.-C. (2003). Biocultural orchestration of developmental plasticity across levels: The interplay of biology and culture in shaping the mind and behavior across the life span. *Psychological Bulletin*, 129(2), 171–194.

Li, Y., Gong, Y., Burmeister, A., Wang, M., Alterman, V., Alonso, A., and Robinson, S. (2020). Leveraging age diversity for organizational performance: An intellectual capital perspective. *Journal of Applied Psychology*. Advance online publication. doi: http://dx.doi.org/10.1037/apl0000497.

Liberman, Z., Woodward, A.L., and Kinzler, K.D. (2017). The origins of social categorization. *Trends in Cognitive Sciences*, 21(7), 556–568.

Lichtman, M. (2014). *Qualitative Research for the Social Sciences*. Sage. doi: https://dx.doi.org/10.4135/9781544307756.

Lippmann, S. (2008). Rethinking risk in the new economy: Age and cohort effects on unemployment and reemployment. *Human Relations*, 61, 1259–1292.

Liptak, A. (2020, January 15). In age bias case, justices discuss 'O.K. Boomer' and eggless cakes. *The New York Times*. Available: https://www.nytimes.com/2020/01/15/us/supreme-court-age-bias-ok-boomer.html?action=click&module=RelatedLinks&pgtype=Article.

Lorenz, T. (2019, October 29). 'OK Boomer' marks the end of friendly generational relations. *The New York Times*. Available: https://www.nytimes.com/2019/10/29/style/ok-boomer.html.

Lorenzo, R., and Reeves, M. (2018). How and where diversity drives financial performance. *Harvard Business Review*. Available: https://hbr.org/2018/01/how-and-where-diversity-drives-financial-performance.

Lovely, S., and Buffum, A.G. (2007). *Generations at School: Building an Age-Friendly Learning Community*. Thousand Oaks, CA: Corwin Press.

Luo, L., Hodges, J., Winship, C., and Powers, D. (2016). The sensitivity of the intrinsic estimator to coding schemes: Comment on Yang, Schulhofer-Wohl, Fu, and Land. *American Journal of Sociology*, 122(3), 930–961.

Lyons, S., and Kuron, L. (2014). Generational differences in the workplace: A review of the evidence and directions for future research. *Journal of Organizational Behavior*, 35(SUPPL.1), S139–S157.

Lyons, B., Alonso, A., Moorman, R., and Miller, A. (2020). Implications for selection. In B. Hoffman, M. Shoss, and L. Wegman (Eds.), *The Cambridge Handbook of the Changing Nature of Work and Workers*. Cambridge: Cambridge University Press.

MacLean, A., and Elder, G.H., Jr. (2007). Military service in the life course. *Annual Review of Sociology*, 33, 175–196.

MaCurdy, T., and Mroz, T. (1995). Measuring macroeconomic shifts in wages from cohort specifications. Unpublished Manuscript, Stanford University and University of North Carolina.

Magnusson, D. (Ed.) (1996). *The Life Span Development of Individuals: Behavioural, Neurobiological and Psychosocial Perspectives*. Cambridge, England: Cambridge University Press.

Mannheim, K. (1952). The problem of generations. In P. Kecskemeti (Ed.), *Karl Mannheim: Essays* (pp. 276–322). New York: Routledge.

Marquis, J.P., Lim, N., Scott, L., Harrell, M.C., and Kavanagh, J. (2008). *Managing Diversity in Corporate America: An Exploratory Analysis*, Santa Monica, CA: RAND Corporation, OP-206-RC, 2008. Available: https://www.rand.org/pubs/occasional_papers/OP206.html.

Maxey, H., Bishop-Josef, S. and Goodman, B. (2018). *Unhealthy and Unprepared. National Security Depends on Promoting Healthy Lifestyles from an Early Age.* Council for a Strong America. Available: https://strongnation.s3.amazonaws.com/documents/484/389765e0-2500-49a2-9a67-5c4a090a215b.pdf?1539616379&inline;%20filename=%22Unhealthy%20and%20Unprepared%20report.pdf%22.

Meacham v. Knolls Atomic Power Laboratory, 554 U.S. 84 (2008). Available: https://supreme.justia.com/cases/federal/us/554/06-1505/index.pdf.

Meriac, J.P., Woehr, D.J., and Banister, C. (2010). Generational differences in work ethic: An examination of measurement equivalence across three cohorts. *Journal of Business and Psychology,* 25(2), 315–324. doi: https://doi.org/10.1007/s10869-010-9164-7.

Meritor Savings Bank, FSB v. Vinson, 477 US 57 (1986). Available: https://tile.loc.gov/storage-services/service/ll/usrep/usrep477/usrep477057/usrep477057.pdf.

Merriam-Webster. (2019). Generation. Merriam-Webster.com. Available: https://www.merriam-webster.com/dictionary/generation.

Mezzofiore, G. (2019, November 7). A 25-year old politician got heckled during a climate crisis speech. Her deadpan retort: 'OK, boomer'. *CNN World.* Available: https://www.cnn.com/2019/11/06/asia/new-zealand-ok-boomer-trnd/index.html.

Milan v. Dediol v. Best Chevrolet, Inc., et al., No. 10-30767 (5th Cir. 2011). Available: https://caselaw.findlaw.com/us-5th-circuit/1579910.html.

Miller, L.J., and Lu, W. (2018). Gen Z is set to outnumber millennials within a year. *Bloomberg Economics.* Available: https://www.bloomberg.com/news/articles/2018-08-20/gen-z-to-outnumber-millennials-within-a-year-demographic-trends.

Mishel, L., Schmitt, J., and Shierholz, H. (2013). *Don't Blame the Robots: Assessing the Job Polarization Explanation of Growing Wage Inequality. EPI-CEPR Working Paper.* Washington, DC: Economic Policy Institute.

Mitchell, S. (1995). *The Official Guide to Generations.* New York: New Strategist Publications, Inc.

———. (1998). *American Generations.* New York: New Strategist Publications, Inc.

Moen, P., Kelly, E., Fan, W., Lee, S-R., Almeida, D., Kossek, E.F., and Buxto, O. (2016a). Does a flexibility/support organizational initiative improve high-tech employees' well-being? Evidence from the work, family, and health network. *American Sociological Review,* 81(1), 134–164.

Moen, P., Kojola, E., Kelly, E.L., and Karakaya, Y. (2016b). Men and women expecting to work longer: Do changing work conditions matter? *Work, Aging and Retirement,* 2, 321–344.

Moen, P., Kelly, E., Oakes, J.M., Lee, S-R., Bray, J., Almeida, D., Hammer, L, Hurtado, D., and Buxto, O. (2017). Can a flexibility/support initiative reduce turnover intentions and exits? Results from the work, family, and health network. *Social Problems,* 64(1), 53–85.

Morgeson, F.P., and Campion, M.A. (2003). Work design. In W.C. Borman, D.R. Ilgen, and R.J. Klimoski (Eds.), *Handbook of Psychology: Industrial and Organizational Psychology* (Vol. 12, pp. 423–452). Hoboken, NJ: John Wiley & Sons.

Mühlig-Versen, A., Bowen, C.E., and Staudinger, U.M. (2012). Personality plasticity in later adulthood: Contextual and personal resources are needed to increase openness to new experiences. *Psychology and Aging,* 27(4), 855.

Muro, M., Liu, S., Whiton, J., and Kulkarni, S. (2017). *Digitalization and the American Workforce.* Washington, DC: Brookings Institution.

Myers, M. (2018). The Army is supposed to be growing, but this year, it didn't at all. *Army Times.* Available: https://www.armytimes.com/news/your-army/2018/09/21/the-army-is-supposed-to-be-growing-but-this-year-it-didnt-at-all.

National Academies of Sciences, Engineering, and Medicine. (2017a). *Building America's Skilled Technical Workforce*. Washington, DC: The National Academies Press. doi: https://doi.org/10.17226/23472.

———. (2017b). *Information Technology and the U.S. Workforce: Where Are We and Where Do We Go from Here?* Washington, DC: The National Academies Press. doi: https://doi.org/10.17226/24649.

———. (2018). *How People Learn II: Learners, Contexts, and Cultures*. Washington, DC: The National Academies Press. doi: https://doi.org/10.17226/24783.

———. (2019). *Strengthening the Military Family Readiness System for a Changing American Society*. Washington, DC: The National Academies Press. doi: https://doi.org/10.17226/25380.

NCES (National Center for Education Statistics). (2018). *National Postsecondary Student Aid Study (NPSAS:2000, NPSAS:12, and NPSAS:16) for 1999–2000, 2011–12, and 2015–16. Table 331.95*. Washington, DC: U.S. Department of Education. Available: https://nces.ed.gov/programs/digest/d18/tables/dt18_331.95.asp.

———. (2019). *The Condition of Education 2019* (NCES 2019-144), Educational Attainment of Young Adults. Washington, DC: U.S. Department of Education. Available: https://nces.ed.gov/fastfacts/display.asp?id=27.

National Restaurant Association. (2019). *Hospitality Industry Turnover Rate Ticked Higher in 2018*. Available: https://restaurant.org/Articles/News/Hospitality-industry-turnover-rate-ticked-higher.

NRC (National Research Council). (1999). *The Changing Nature of Work: Implications for Occupational Analysis*. Washington, DC: The National Academies Press. doi: https://doi.org/10.17226/9600.

———. (2002). *Letter Report from the Committee on Youth Population and Military Recruitment*. Washington, DC: The National Academies Press. doi: https://doi.org/10.17226/10317.

Nelson, E.A., and Dannefer, D. (1992). Aged heterogeneity: Fact or fiction? The fate of diversity in gerontological research. *The Gerontologist*, 32(1), 17–23.

New Development Bank. (2017). *The Role of BRICS in the World Economy and International Development*. Available: https://reddytoread.files.wordpress.com/2017/09/brics-2017.pdf.

Ng, T.W., and Feldman, D.C. (2008). The relationship of age to ten dimensions of job performance. *Journal of Applied Psychology*, 93(2), 392–423.

———. (2010a). Organizational tenure and job performance. *Journal of Management*, 36(5), 1220–1250.

———. (2010b). The relationships of age with job attitudes: A meta-analysis. *Personnel Psychology*, 63, 677–718.

North, M.S. (2019). A GATE to understanding "older" workers: Generation, age, tenure, experience. *Academy of Management Annals*, 13(2), 414–443. doi: https://doi.org/10.5465/annals.2017.0125.

North, M.S., and Shakeri, A. (2019). Workplace subjective age multidimensionality: Generation, age, tenure, experience (GATE). *Work, Aging and Retirement*, 5, 281–286.

Ogborn, M. (1998). *Spaces of Modernity: London's Geographies, 1680–1780*. London: Guilford Press.

Osterman, P. (1999). *Securing Prosperity: The American Labor Market: How It Has Changed and What to Do About It*. Princeton, NJ: Princeton University Press.

Osterman, P., Kochan, T.A., Locke, R.M., and Piore, M.J. (2002). *Working in America: A Blueprint for the New Labor Market*. Cambridge, MA: MIT Press.

Parker, K., Horowitz, J.M., Brown, A., Fry, R., Cohn, D., and Igielnik, R. (2018). Demographic and economic trends in urban, suburban, and rural communities. *Social and Demographic Trends*. Washington, DC: Pew Research Center. Available: https://www.pewsocialtrends.org/2018/05/22/demographic-and-economic-trends-in-urban-suburban-and-rural-communities.

Parry, E., and McCarthy, J. (2017). *The Palgrave Handbook of Age Diversity and Work*. United Kingdom: Palgrave Macmillan.

Parry, E., and Urwin, P. (2011). Generational differences in work values: A review of theory and evidence. *International Journal of Management Reviews* 13(1), 79–96. doi: https://doi.org/10.1111/j.1468-2370.2010.00285.x.

———. (2017). The evidence base for generational differences: Where do we go from here? *Work, Aging, and Retirement* 3(2), 140–148. doi: https://doi.org/10.1093/workar/waw037.

Pawlyk, O. (2019). *Targeted Messaging: Military Recruiters Getting Creative to Reach Gen Z*. Available: https://www.military.com/daily-news/2019/02/24/targeted-messaging-military-recruiters-getting-creative-reach-gen-z.html.

Pedhazur, E.J., and Pedhazur Schmelkin, L. (1991). *Measurement, Design and Analysis: An Integrated Approach*. Hillsdale, NJ: Lawrence Erlbaum Associates.

Perry, E.L., Golom, F.D., Catenacci, L., Ingraham, M.E., Covais, E.M., and Molina, J.J. (2017). Talkin' 'bout your generation: The impact of applicant age and generation on hiring related perceptions and outcomes. *Work, Aging and Retirement*, 3, 186–199.

Petrosky-Nadeau, N., and Valletta, R.G. (2019). Unemployment: Lower or longer? *FRBSF Economic Letter*, Federal Reserve Bank of San Francisco. Available: https://www.frbsf.org/economic-research/publications/economic-letter/2019/august/unemployment-lower-for-longer/.

Petty, G.C. (2013). The perceived work ethic of K-12 teachers by generational status: Generation X vs. baby boom generation. *International Journal of Adult Vocational Education and Technology*, 4(2), 54–65.

Pitts, J.M. (2020, May 14). Shortage in the nursing field amid pandemic is causing concern. *Washington Post*. Available: https://www.washingtonpost.com/local/shortage-in-the-nursing-field-amid-pandemic-is-causing-concern/2020/05/14/24890a44-93d9-11ea-82b4-c8db161ff6e5_story.html.

Posthuma, R.A., and Campion, M. (2009). Age stereotypes in the workplace: Common stereotypes, moderators, and future research directions. *Journal of Management*, 35(1), 158–188.

PRC (Pew Research Center). (2013). *Breadwinner Moms*. Available: https://www.pewsocialtrends.org/wp-content/uploads/sites/3/2013/05/Breadwinner_moms_final.pdf.

———. (2015). *Raising Kids and Running a Household: How Working Parents Share the Load*. Available: https://www.pewsocialtrends.org/2015/11/04/raising-kids-and-running-a-household-how-working-parents-share-the-load.

Protzko, J., and Schooler, J.W. (2019). Kids these days: Why the youth of today seem lacking. *Science Advances*, 5(10), eaav5916.

Rabin v. PricewaterhouseCoopers LLP, 236 F.Supp.3d 1126 (N.D. Cal. 2017). Available: https://casetext.com/case/rabin-v-pricewaterhousecoopers-llp.

Raphelson, S. (2014, October 6). From GIs to Gen Z (or is it iGen?): How generations get nicknames. *National Public Radio*. Available: https://www.npr.org/2014/10/06/349316543/don-t-label-me-origins-of-generational-names-and-why-we-use-them.

Reid, K.C. (2018, March 20). How the network generation is changing the millennial military. Commentary. *War on the Rocks*. Available: https://warontherocks.com/2018/03/how-the-network-generation-is-changing-the-millennial-military.

Richard, O.C., and Miller, C.D. (2013). Considering diversity as a source of competitive advantage in organizations. In Roberson, Q.M. (Ed.), *The Oxford Handbook of Diversity and Work* (pp. 1–21). Oxford, UK: Oxford University Press.

Riffkin, R. (2014). *Average U.S. Retirement Age Rises to 62.* Gallup. Available: https://news.gallup.com/poll/168707/average-retirement-age-rises.aspx.

Riley, M.W. (1973). Aging and cohort succession: Interpretations and misinterpretations. *Public Opinion Quarterly*, 37(1), 35–49.

———. (1987). On the significance of age in sociology. *American Sociological Review* 52(1), 1. doi: https://doi.org/10.2307/2095388.

Ritchie, H., and Roser, M. (2020). *Urbanization.* Available: https://ourworldindata.org/urbanization.

Roberts, B.W., Walton, K.E., and Viechtbauer, W. (2006). Patterns of mean-level change in personality traits across the life course: A meta-analysis of longitudinal studies. *Psychological Bulletin*, 132(1), 1–25.

Rones, P.L., Ig, R.E., and Gardner, J.M. (1997). Trends in hours of work since the mid-1970s. *Monthly Labor Review.* U.S. Bureau of Labor Statistics. Available: https://www.bls.gov/opub/mlr/1997/04/art1full.pdf.

Rosch, E. (1978). Principles of categorization. In E. Rosch and B.B. Lloyd (Eds.), *Cognition and Categorization.* Hillsdale, NJ: Erlbaum.

Rosenberg, E.. and Long, H. (2020, June 6). Unemployment rate drops and 2.5 million jobs added, after states reopened. *The Washington Post.* Available: https://www.washingtonpost.com/business/2020/06/05/may-2020-jobs-report/.

Rosenfeld, J. (2014). *What Unions No Longer Do.* Cambridge, MA: Harvard University Press.

Roser, M. (2020a). *Employment in Agriculture.* Available: https://ourworldindata.org/employment-in-agriculture.

———. (2020b). *Working Hours.* Available: https://ourworldindata.org/working-hours.

Ross, W.D. (Ed.) (2010). *Rhetoric.* New York: Cosimo, Incorporated.

Rotundo, M., and Sackett, P. (2002). The relative importance of task, citizenship, and counterproductive performance to global ratings of job performance: A policy-capturing approach. *Journal of Applied Psychology*, 87(1), 66–80.

Rudolph, C.W., and Zacher, H. (2015). Intergenerational perceptions and conflicts in multi-age and multigenerational work environments. In *Facing the Challenges of a Multi-Age Workforce: A Use-Inspired Approach* (pp. 253–282). New York: Routledge.

———. (2017). Considering generations from a lifespan developmental perspective. *Work, Aging, and Retirement* 3:113–129. doi: https://doi.org/10.1093/workar/waw019.

Rudolph, C.W., Rauvola, R.S., and Zacher, H. (2017). Leadership and generations at work: A critical review. *Leadership Quarterly*, 29(1), 44–57. doi: https://doi.org/10.1016/j.leaqua.2017.09.004.

———. (2018). Leadership and generations at work: A critical review. *Leadership Quarterly*, 29(1), 44–57.

Ryder, N.B. (1965). The cohort as a concept in the study of social change. *American Sociological Review*, 30(6), 843–861.

Sakdiyakorn, M., and Wattanacharoensil, W. (2018). Generational diversity in the workplace: A systematic review in the hospitality context. *Cornell Hospitality Quarterly*, 59(2), 135–159.

Salas, E., Stagl, K.C., and Burke, C.S. (2004). 25 years of team effectiveness in organizations: Research themes and emerging needs. In C.L. Cooper, and I.T Robertson (Eds.), *International Review of Industrial and Organizational Psychology* (Vol. 19, pp. 47–91). Chichester, UK: John Wiley & Sons.

Schmidt, F.L., and Hunter, J.E. (1998). The validity and utility of selection methods in personnel psychology: Practical and theoretical implications of 85 years of research findings. *Psychological Bulletin*, 124(2), 262–274.

Schmitt, N., and Kuljanin, G. (2008). Measurement invariance: Review of practice and implications. *Human Resource Management Review* 18(4), 210–222.

Schwartz, K. (2019, September 19). Thriving hotel industry scrambles for workers. *New York Times*. Available: https://www.nytimes.com/2019/09/09/travel/hotel-jobs.html.

Scopelliti, D. (2014). *Middle-Skill Jobs Decline as U.S. Labor Market Becomes More Polarized*. Washington, DC: U.S. Bureau of Labor Statistics.

Segal, M.W. (1986). The military and the family as greedy institutions. *Armed Forces and Society*, 13(1), 9–38.

Shaftesbury, A.A.C. (1868). *Speeches of the Earl of Shaftesbury, Upon Subjects Having Relation Chiefly to the Claims and Interests of the Labouring Class*. London: Chapman and Hall.

Shepherd, W. (2014). The heterogeneity of well-being: Implications for HR management practices. *Industrial and Organizational Psychology*, 7, 579–583. doi: https://doi.org/10.1017/s1754942600006933.

Shore, L.M., Cleveland, J.M., and Sanchez, D. (2018). Inclusive workplaces: A review and model. *Human Resource Management Review*, 28(2), 176–189.

Smith v. City of Jackson, Miss., 544 U.S. 228 (2005). Available: https://supreme.justia.com/cases/federal/us/544/03-1160/index.pdf.

Stafford, D.E., and Griffis, H.S. (2008). *A Review of Millennial Generation Characteristics and Military Workforce Implications*. Alexandria, VA: The CNA Corporation. Available: https://www.cna.org/CNA_files/PDF/D0018211.A1.pdf.

Stanford Center on Poverty and Inequality. (2011). *20 Facts about U.S. Inequality that Everyone Should Know*. Available: https://inequality.stanford.edu/publications/20-facts-about-us-inequality-everyone-should-know.

Stimpfel, A.W., and Dickson, V.V. (2019). Working across generations to boost staff nurse retention. *Western Journal of Nursing Research*, 42(6), 395–396.

Stimpfel, A.W., Arabadjian, M., Liang, E., Sheikhzadeh, A., Weiner, S.S., Dickson, V.V. (2019). Organization of work factors associated with work ability among aging nurses. *Western Journal of Nursing Research*, 42(6), 397–404. doi: https://doi.org/10.1177/0193945919866218.

Strauss, W., and Howe, N. (1991). *Generations: The History of America's Future 1584–2069*. New York: Morrow, William and Co.

———. (1998). *The Fourth Turning: An American Prophecy*. New York: Broadway Books.

Stuart, A. (1968). Sample surveys. II: Nonprobability sampling. *International Encyclopaedia of the Social Sciences* 13, 612–616.

Sullivan, K. (2016). Workforce growth in community-based care: Meeting the needs of an aging population. *Monthly Labor Review*. U.S. Bureau of Labor Statistics. doi: https://doi.org/10.21916/mlr.2016.53.

Tam, M., Korczynski, M., and Frenkel, S. (2002) Organizational and occupational commitment: Knowledge workers in large corporations. *Journal of Management Studies*, 39(6), 775–801.

Tannenbaum, S.I., Beard, R.L., McNall, L.A., and Salas, E. (2010). Informal learning and development in organizations. In S.W.J. Kozlowski and E. Salas (Eds.), *Learning, Training, and Development in Organizations* (pp. 303–331). New York: Routledge/Taylor and Francis Group.

Tay, L., Meade, A.W., and Cao, M. (2015). An overview and practical guide to IRT measurement equivalence analysis. *Organizational Research Methods* 18(1), 3–46.

Thie, H.J., Harrell, M.C., and Thibault, M. (2003). *Officer Sabbaticals: Analysis of Extended Leave Options*. National Defense Research Institute. Santa Monica, CA: RAND.

Toosi, M. (2002). A century of change: The U.S. labor force, 1950–2050. *Monthly Labor Review*. Washington, DC: U.S. Bureau of Labor Statistics. Available: https://www.bls.gov/opub/mlr/2002/05/art2full.pdf.

Tracey, J.B., Tannenbaum, S.I., and Kavanaugh, M.J. (1995). Applying trained skills on the job: The importance of the work environment. *Journal of Applied Psychology*, 80, 239–252. doi: http://dx.doi.org/10.1037/0021-9010.80.2.239.

Truxillo, D.M., Cadiz, D.M., and Hammer, L.B. (2015). Supporting the aging workforce: A review and recommendations for workplace intervention research. *Annual Review of Organizational Psychology and Organizational Behavior*, 2(1), 351–381.

Turban, D., and Greening, D. (1997). Corporate social performance and organizational attractiveness to prospective employees. *The Academy of Management Journal*, 40(3), 658–672.

Turban, S., Wu, D., Zhang, L. (2019). When gender diversity makes firms more productive. *Harvard Business Review*. Available: https://hbr.org/2019/02/research-when-gender-diversity-makes-firms-more-productive.

Twenge, J.M. (2006). *Generation Me: Why Today's Young Americans Are More Confident, Assertive, Entitled—and More Miserable Than Ever Before*. New York: Free Press.

———. (2010). A review of the empirical evidence on generational differences in work attitudes. *Journal of Business and Psychology* 25(2), 201–210.

———. (2018). *iGen*. New York: Atria Books.

Twenge, J.M., and Campbell, W.K. (2001). Age and birth cohort differences in self-esteem: A cross-temporal meta-analysis. *Personality and Social Psychology Review*, 5, 321–344.

Twenge, J.M., and Campbell, S.M. (2008). Generational differences in psychological traits and their impact on the workplace. *Journal of Managerial Psychology* 23(8), 862–877.

Twenge, J.M., Campbell, W.K., and Freeman, E.C. (2012). Generational differences in young adults' life goals, concern for others, and civic orientation, 1966–2009. *Journal of Personality and Social Psychology*, 102(5), 1045–1062.

Twenge, J.M., Carter, N.T., and Campbell, W.K. (2017). Age, time period, and birth cohort differences in self-esteem: Reexamining a cohort-sequential longitudinal study. *Journal of Personality and Social Psychology*, 112(5), e9–e17. doi: https://doi.org/10.1037/pspp0000122.

Twenge, J.M., Campbell, S.M., Hoffman, B.J., and Lance, C.E. (2010). Generational differences in work values: Leisure and extrinsic values increasing, social and intrinsic values decreasing. *Journal of Management* 36(5), 1117–1142. doi: https://doi.org/10.1177/0149206309352246.

Urick, M.J., Hollensbe, E.C., Masterson, S.S., and Lyons, S.T. (2017). Understanding and managing intergenerational conflict: An examination of influences and strategies. *Work, Aging and Retirement*, 3, 166–185.

U.S. Army. (2019). *The Army People Strategy*. Available: https://www.army.mil/e2/downloads/rv7/the_army_people_strategy_2019_10_11_signed_final.pdf.

U.S. Census Bureau. (2010). Employment status of married-couple families by presence of own children under 18 years: 2008–2009. *Release Number ACSBR/09-10*. Available: https://www2.census.gov/library/publications/2010/acs/acsbr09-10.pdf.

U.S. Census Bureau. (2015). Millennials outnumber baby boomers and are far more diverse. *Census Bureau Reports. Release Number CB15-113*. Available: https://www.census.gov/newsroom/press-releases/2015/cb15-113.html.

Vandenberg, R.J., and Lance, C.E. (2000). A review and synthesis of the measurement invariance literature: Suggestions, practices, and recommendations for organizational research. *Organizational Research Methods* 3(1), 4–70.

Villarreal v. R.J. Reynolds Tobacco Co., 839 F.3d 958 (2016). Available: http://media.ca11. uscourts.gov/opinions/pub/files/201510602.enb.pdf.

Wall Street Journal. (2017, December 4). *Millennials. Style & Substance, 30(11).* Available: https://blogs.wsj.com/styleandsubstance/2017/12/04/vol-30-no-11-millennials.

Walker, T. (2019, March). Five key trends in the teacher workforce. *NEA Today.* Available: http://neatoday.org/2019/03/13/5-trends-in-the-teaching-profession.

Wang, H.L. (2020, April 23). 'I hear the agony': Coronavirus crisis takes toll on NYC's first responders. *National Public Radio.* Available: https://www.npr.org/2020/04/23/842011186/i-hear-the-agony-coronavirus-crisis-takes-toll-on-nyc-s-first-responders.

Wang, M., Olson, D., and Shultz, K. (2013). *Mid and Late Career Issues: An Integrative Perspective.* New York: Psychology Press.

Wegman, L.A., Hoffman, B.J., Carter, N.T., Twenge, J.M., and Guenole, N. (2018). Placing job characteristics in context: Cross-temporal meta-analysis of changes in job characteristics since 1975. *Journal of Management* 44, 352–386.

Weil, D. (2014). *The Fissured Workplace: Why Work Became so Bad for so Many and What Can be Done to Improve It.* Cambridge, MA: Harvard University Press.

Weiss, D., and Lang, F.R. (2009). Thinking about my generation: Adaptive effects of a dual age identity in later adulthood. *Psychology and Aging,* 24, 729–734.

———. (2012). Two faces of age identity. *The Journal of Gerontopsychology and Geriatric Psychiatry,* 25, 5–14.

Weiss, D., and Perry, E.L. (2020). Implications of generational and age metastereotypes for older adults at work: The role of agency, stereotype threat, and job search self-efficacy. *Work, Aging and Retirement,* 6, 15–27.

Wenger, J.B., Knapp, D., Mahajan, P., Orvis, B.R., and Tsai, T. (2019). *Developing a national recruiting difficulty index (RR-2637-A).* Santa Monica, CA: RAND Corporation. Available: https://www.rand.org/pubs/research_reports/RR2637.html.

Wertz, F.J. (2014). Qualitative inquiry in the history of psychology. *Qualitative Psychology* 1, 4–16. doi: http://dx.doi.org/10.1037/qup0000007.

Western Air Lines, Inc. v. Criswell, 472 U.S. 400, 422 (1985). Available: https://tile.loc.gov/storage-services/service/ll/usrep/usrep472/usrep472400/usrep472400.pdf.

Wieck K.L., Dols, J., and Landrum, P. (2010). Retention priorities for the intergenerational nurse workforce. *Nursing Forum,* 45(1), 7–17.

Williams, G. (2019). Management millennialism: Designing the new generation of employee. *Work, Employment, and Society,* 1–17.

Winship, C., and Harding, D.J. (2008). A mechanism-based approach to the identification of age-period-cohort models. *Sociological Methods and Research,* 36(3), 362–401. doi: https://doi.org/10.1177/0049124107310635.

Wong, L. (2000). *Generations Apart: Xers and Boomers in the Officer Corps.* Carlisle, PA: Strategic Studies Institute, U.S. Army War College.

Work Institute. (2019). *2019 Retention Report.* Available: https://info.workinstitute.com/hubfs/2019%20Retention%20Report/Work%20Institute%202019%20Retention%20Report%20final-1.pdf.

Workgroup on the Early Childhood Workforce and Professional Development (2016). *Proposed Revisions to the Definitions for the Early Childhood Workforce in the Standard Occupational Classification: White paper commissioned by the Administration for Children and Families, U.S. Department of Health and Human Services (OPRE Report 2016-45).* Washington, DC: Office of Planning, Research and Evaluation, Administration for Children and Families, U.S. Department of Health and Human Services.

World Health Organization. (2018). *Ageing and health. Fact Sheet.* Available: https://www.who.int/news-room/fact-sheets/detail/ageing-and-health.

Wright, A.D. (2018). *Employers Say Accommodating Millennials is a Business Imperative.* Society of Human Resources Management. Available: https://www.shrm.org/resourcesandtools/hr-topics/employee-relations/pages/accommodating-millennials-is-a-business-imperative.aspx.

Wylie, R. (2017). *New Firefighters Require New Approaches.* Available: https://www.firerescue1.com/firefighter-training/articles/new-firefighters-require-new-approaches-sJkpErbD1e1HdgKn.

Yang, Y., and Land, K.C. (2013). *Age-Period-Cohort Analysis: New Models, Methods, and Empirical Applications.* Boca Raton, FL: CRC Press.

Yoong, P., and Huff, S. (2006). *Managing IT Professionals in the Internet Age.* Hershey, PA: Idea Group Publishing.

Yoshida, K., and Keene, D. (1998). *Essays in Idleness: The Tsurezuregusa of Kenk .* United Kingdom: Columbia University Press.

Zemke, R., Raines, C., and Filipczak, R. (1999). *Generations at Work: Managing the Clash of Veterans, Boomers, Xers, and Nexters in Your Workplace.* New York: American Management Association.

Zhang, L. (2020). An institutional approach to gender diversity and firm performance. *Organization Science*, 31(2), 245–534.

Appendix A

Details of Literature Review

This appendix describes the committee's strategy for gathering and reviewing the business management and behavioral science literature on generational attitudes and behaviors in workforce management and employment practices. The committee's primary objective was to identify and take stock of this body of literature: its size, the types of research questions examined, the types of research designs, and any agreement on findings among researchers. Our initial search of the literature uncovered a number of literature reviews that had already been conducted on this body of work, and our focus turned to understanding what these reviews had found. It was clear that there was much debate on the quality and value of research in this area. We found that where efforts had been made to synthesize findings across studies, the conclusions drawn were inconsistent, and there was disagreement on whether effect sizes on "generation effects" were significant enough to be meaningful and whether observed effects were even related to generations or had other explanations. The committee's findings and conclusions that resulted from looking at the debates in the literature are discussed in Chapter 4. This appendix outlines the particular articles the committee reviewed and gives readers a sense of where the literature can be found, what topics are covered, and what primary research designs were used.

Through the National Academies Research Center, the committee conducted electronic searches in Scopus and ProQuest (see the search syntax in Box A-1). We supplemented our electronic searches with suggestions made by committee members and invited presenters and citations of relevant articles in the previously published literature reviews on this topic identified

BOX A-1
Database Search Syntax

Scopus:

TITLE-ABS-KEY(("generational difference*" OR "generational effect*" OR "age effect*" OR "period effect*" OR "cohort effect*") AND (employment* OR occupation* OR "at work" OR workforce* OR job OR workplace*) AND ("Analytical method*" OR "longitudinal study" OR "longitudinal studies" OR "longitudinal survey" OR "longitudinal data" OR "observational study" OR "observational studies" OR "cohort study" OR "cohort studies" OR "survey research" OR "survey method" OR Questionnaire* OR Interview* OR "empirical research")) AND PUBYEAR > 1979

TITLE-ABS-KEY(("Generational difference" OR "generational differences" OR ("baby boomer*" AND ("generation y" OR "generation z" OR "millennial generation" OR "generation me" OR "igen")) AND ({a total of} OR {N=} OR {survey of} OR data OR empirical OR cohort*) AND ("meaningful work" OR workforce OR workplace OR employment OR "job satisfaction" OR "work values" OR "work life" OR "employee engagement")) AND PUBYEAR > 1979

ProQuest Research Library:

ti(("Generational difference" OR "generational differences")) AND noft(("Analytical method*" OR "longitudinal study" OR "longitudinal studies" OR "longitudinal survey" OR "longitudinal data" OR "observational study" OR "observational studies" OR "cohort study" OR "cohort studies" OR "survey research" OR "survey method" OR Questionnaire* OR Interview* OR "empirical research")) AND noft(("meaningful work" OR workforce OR workplace OR employment OR "job satisfaction" OR "work values" OR "work life" OR "employee engagement"))

su(generations) AND noft((("Generational difference" OR "generational differences"))) AND noft((("Analytical method*" OR "longitudinal study" OR "longitudinal studies" OR "longitudinal survey" OR "longitudinal data" OR "observational study" OR "observational studies" OR "cohort study" OR "cohort studies" OR "survey research" OR "survey method" OR Questionnaire* OR Interview* OR "empirical research" or data or empirical or cohort* OR survey))) AND noft((employment* OR occupation* OR "at work" OR workforce* OR job OR workplace*))

NOTE: These databases were selected by the National Academies Research Center because they are comprehensive, multidisciplinary, accessible, and available to Academies staff and their committees. Scopus is one of the largest multidisciplinary abstract databases containing peer-reviewed literature. ProQuest Research Library nicely complements Scopus as it also targets multidisciplinary peer-reviewed literature, and it includes trade publications and magazines to round out the search.

during our search. We classified a long list of references by their research designs. Further, we reviewed and discussed the observations, findings, and conclusions of previously published literature reviews. Additionally, we conducted a small pilot review of a subset of the articles identified in our electronic search to appreciate the issues discussed in earlier critiques of this literature.

LITERATURE SEARCH

We conducted the first electronic search at the beginning of the study (March 2019) to identify articles published after 1979 in the United States and internationally. This search resulted in 306 articles (96% of which were published after 1999). Of these, we found 57 to be irrelevant to our study or duplicative. To ensure that we considered other work-related articles on generational attitudes and behaviors without attention to differences, we conducted a second electronic search in August 2019 for articles published after 1999 using the same databases and similar search syntax, but without the search terms for "differences" or "effect." This second search resulted in another 121 articles.

Among the articles were 15 previously published literature reviews on this same body of research, each of which reflects a different approach. We categorized the reviews into four types: (1) meta-analyses (reviews that quantitatively compare findings across studies by calculating the effect sizes or other metric to quantify the relationship between generational membership or age- and work-related outcomes, values, or attitudes); (2) systematic or structured reviews (reviews that descriptively identify and synthesize information about findings across empirical studies); (3) sector-specific reviews (reviews that descriptively identify and synthesize information about findings from empirical studies focused on specific employment sectors); and (4) commentaries and methodological validations (articles that explore methodological, analytical, and/or theoretical issues in the literature). Table A-1 summarizes the methods, major findings, and conclusions from these reviews. An additional 16 articles were also flagged as literature reviews, as opposed to empirical studies, but the authors of these articles were less systematic or structured in their reviews relative to the reviews described in Table A-1, which were designed to reflect the state of the evidence.

To finalize our list of generational literature, we compared our initial list with articles identified by other reviews, recording another 188 articles. We screened the titles and abstracts of these articles to ensure that they focused on generational attitudes and behaviors in the workforce and to determine their research designs where possible. In some cases, we relied on the assessment of other reviews to determine research designs, while in other cases we screened the full articles.

TABLE A-1 Summary of Methods, Major Findings, and Conclusions from Meta-Analyses and Structured Reviews

Sources	Methods, Major Findings, Conclusions
	Meta-Analyses
1. Ng, T.W.H., and Feldman, D.C. (2010). The relationships of age with job attitudes: A meta-analysis, *Personnel Psychology*, 63, 677–718.	• This meta-analysis focuses on the relationship of age as opposed to generation with job attitudes. It is an indication of how much more research is available on age and work. The review also touches on questions related to generational differences and provides references to the generational literature. • Job attitudes are defined as summary evaluations of psychological objects in the work domain in three broad categories: task-based attitudes, people-based attitudes, and organization-based attitudes. • Studies were conducted in the 1970s to 2009. • Most studies are cross-sectional. • Studies used a standard protocol with meta-analytic correlations measuring the relationship between age and attitudes. • "Results of meta-analyses from more than 800 articles indicate that the relationships between chronological age and favorable attitudes (and/or to less unfavorable attitudes) toward work tasks, colleagues and supervisors, and organizations are generally significant and weak to moderate in strength. Moderator analyses also revealed that organizational tenure, race, gender, education level, and publication year of study moderate the relationships between age and job attitudes" (p. 677).
2. Costanza, D.P., Badger, J.M., Fraser, R.L., Severt, J.B., and Gade, P.A. (2012). Generational differences in work-related attitudes: A meta-analysis. *Journal of Business and Psychology*, 27(4), 375–394.	• The meta-analysis covers generational differences in three work-related attitude areas: job satisfaction, organizational commitment, and intent to turn over. • Review of published and unpublished research found 20 studies that met inclusion criteria and contained sufficient information to calculate effect sizes, allowing for generational comparisons across four generations (traditionals, baby boomers, generation Xers, and millennials) on these outcomes using 19,961 total subjects. • Studies were conducted between 1995 and 2009. Four of the studies were conducted outside the United States. All studies are cross-sectional. • The paper includes a table of all studies with effect sizes and study characteristics. • "The pattern of results indicates that the relationships between generational membership and work-related outcomes are moderate to small, essentially zero in many cases" (p. 375). Country was not an important factor.

TABLE A-1 Continued

Sources	Methods, Major Findings, Conclusions
3. Jin, J., and Rounds, J. (2012). Stability and change in work values: A meta-analysis of longitudinal studies. *Journal of Vocational Behavior*, 80(2), 326–339.	• Reviews longitudinal studies to investigate stability and change in work values across the life span. • Includes 22 studies that met inclusion criteria. • Uses four age categories collapsed into two: baby boomers (boomers; born 1946–1964) and generation X (genX; born 1965–1981). • Among other results, "with regard to generational difference, boomers and genX differed little in terms of intrinsic values. However, boomers increased their extrinsic values over time (d = .06), while those of genX decreased at a similar level (d = −.07).[a] However, neither was significantly different from zero. For both boomers and genX, social values decreased significantly over time, with the magnitude of decrease for genX (d = −.16) larger than that for boomers (d = −.12). With respect to status values, those of boomers remained unchanged over time, while those of genX decreased dramatically (d = −.13); however, the confidence intervals were 0 for both boomers and genX" (p. 335).[b] • "Consistent with [their] hypothesis, the authors found that while work values remained rather stable when indexed by rank-order stability, they did change when viewed from the mean-level perspective" (p. 335).

	Systematic or Structured Reviews
4. Twenge, J.M. (2010). A review of the empirical evidence on generational differences in work attitudes. *Journal of Business and Psychology*, 25(2), 201–210.	• Reviews the available evidence—primarily papers published in peer-reviewed journals—on generational differences in work values (in the categories of work ethic, work centrality, and leisure; altruistic values; extrinsic versus intrinsic values; affiliation or social values; and job satisfaction and intention to leave) and on personality differences relevant to the workplace. • The studies reviewed used time-lag (which can separate generation from age/career stage) and cross-sectional (which cannot) methods. • The studies reviewed used respondents from Australia, Belgium, Europe as a whole, New Zealand, and the United States. • Where possible, effect sizes are noted for generational differences in terms of d. • "Most studies, including the few time-lag studies, show that GenX and especially 'GenMe' [author's term for genY or millennials] rate work as less central to their lives, value leisure more, and express a weaker work ethic than Boomers and Silents. Extrinsic work values (e.g., salary) are higher in GenMe and especially GenX. Contrary to popular conceptions, there were no generational differences in altruistic values (e.g., wanting to help others). Conflicting results appeared in desire for job stability, intrinsic values (e.g., meaning), and social/affiliative values (e.g., making friends). GenX, and especially GenMe are consistently higher in individualistic traits" (p. 201). • Overall, the author believes that generational differences are important where they appear, as even small changes at the average mean that twice or three times as many individuals score at the top of the distribution.

TABLE A-1 Continued

Sources	Methods, Major Findings, Conclusions
5. Parry, E., and Urwin, P. (2011). Generational differences in work values: A review of theory and evidence. *International Journal of Management Reviews*, 13(1), 79–96.	• "This paper presents a critical review of the theoretical basis and empirical evidence for the popular practitioner idea that there are generational differences in work values" (p. 79). • Reviews literature since 1983 that relates to Western democracies, mainly the United States and Europe but also other parts of the world, to examine whether generations differ within different countries and cultures. • "The concept of generations has a strong basis in sociological theory, but the academic empirical evidence for generational differences in work values is, at best, mixed. Many studies are unable to find the predicted differences in work values, and those that do often fail to distinguish between 'generation' and 'age' as possible drivers of such observed differences. In addition, the empirical literature is fraught with methodological limitations through the use of cross-sectional research designs in most studies, confusion about the definition of a generation as opposed to a cohort, and a lack of consideration for differences in national context, gender and ethnicity" (p. 79).
6. Lyons, S., and Kuron, L. (2014). Generational differences in the workplace: A review of the evidence and directions for future research. *Journal of Organizational Behavior*, 35(SUPPL.1), S139–S157.	• Provides a "critical review" of the research evidence concerning generational differences in a variety of work-related variables, including personality, work values, work attitudes, leadership, teamwork, work–life balance, and career patterns. • Presents longitudinal and cross-sectional evidence. • Comments on the degree to which context (e.g., historical, cultural, occupational, and organizational) is considered in the research. Distinguishes U.S. studies from those conducted elsewhere. • Describes broader generational trends in each area, rather than pairwise comparisons. • Authors indicate that the "growing body of research, particularly in the past 5 years, remains largely descriptive, rather than exploring the theoretical underpinnings of the generation construct. Evidence to date is fractured, contradictory, and fraught with methodological inconsistencies that make generalizations difficult. The results of time-lag, cross-temporal meta-analytic, and cross-sectional studies provide sufficient 'proof of concept' for generation as a workplace variable" (p. S139).

TABLE A-1 Continued

Sources	Methods, Major Findings, Conclusions
7. Woodward, I., Vongswasdi, P., and More, E. (2015). Generational diversity at work: A systematic review of the research. *Working Paper Series,* 2015/48/OB. INSEAD, The Business School for the World.	• This is not a peer-reviewed article but was identified at a committee meeting as a study that was conducted by an international business school and includes a number of references to the generational literature. • The review is very descriptive. The authors claim it is the only systematic review conducted to date. • Reviews 50 studies, finding numerous differences among generations. • The authors conclude that "taken collectively, the findings provide sufficient support for the notion that generational differences are a valid and legitimate form of diversity in organizations. Overall, this empirical evidence suggests that although generations do share certain similarities (with some mixed results that are anything but conclusive), they also differ in various aspects ranging from work values and work attitudes to other work-related preferences and behaviors" (p. 42).
8. Ng, E.S., and Parry, E. (2016). Multigenerational research in human resource management. *Research in Personnel and Human Resources Management*, 34, 1–41.	• Reviews "evidence from existing research studies to establish the areas of differences that may exist among the different generations" (p. 1), with a particular emphasis on the millennial generation. The review strategy is not clearly described. • Describes differences found across studies in personality, work values, psychological contracts, and generational differences that are relevant to human resource management practices in such areas as new workforce entrants, retaining baby boomers, the changing nature of work and careers, the quest for work–life balance, and leadership preferences. • Although the authors recognize that "critics argue that the effect sizes in the differences are small...[they also recognize] sufficient research studies point to meaningful and material differences across the four generations with respect to their work values, attitudes, and career expectations..." (p. 26).
9. Stassen, L., Anseel, F., and Levecque, K. (2016). Generational differences in the workplace: A systematic analysis of a myth. *Gedrag & Organisatie*, 29(1).	• This article had to be translated from the German. • This is a systematic review of empirical studies, providing an overview of the evidence for generational differences in the workplace. • The authors critique 6 randomly chosen cross-sectional studies, then provide a more thorough review of 20 empirical studies from 2005 to 2014, using the following inclusion criteria: (1) the study had to investigate a difference between generations, (2) one of the generations had to be "generation Y," and (3) the study had to have workplace attitude as a dependent variable or value. • The authors conclude that there is little evidence to date in the scientific literature for distinguishing generations with respect to the workplace.

TABLE A-1 Continued

Sources	Methods, Major Findings, Conclusions
10. Rudolph, C.W., Rauvola, R., and Zacher, H. (2018). Leadership and generations at work: A critical review. *Leadership Quarterly*, 29(1), 44–57.	• Presents "a critical review of theory, empirical research, and practical applications regarding generational differences in leadership phenomena" (p. 44). The authors "call for a moratorium to be placed upon the application of the ideas of generations and generational differences to leadership theory, research, and practice" (p. 44). • The authors conducted a structured search only for empirical studies published in peer-reviewed journals to identify the literature relevant to leadership and generations. • They included 18 articles that used cross-sectional and mixed methods, with samples from a variety of industry sectors. • The review found "relatively little empirical research that studies leadership and generations, suggesting that most of the popular literature that claims evidence for generational differences in leadership phenomena is based on little more than (theoretical) supposition and (anecdotal) conjecture" (p. 48). • "Results of cross-sectional survey studies on leadership and generations provide mixed results regarding the existence of generational differences in leadership preferences" (p. 51). • "…results of mixed-method studies on leadership and generations are equivocal in nature: while some qualitative differences between generational cohorts were found, particularly in terms of leadership trait rankings, many of these thematic divergences are not mutually exclusive… and suggest overlap between generations" (p. 52).
	Sector-Specific Reviews
11. Sakdiyakorn, M., and Wattanacharoensil, W. (2018). Generational diversity in the workplace: A systematic review in the hospitality context. *Cornell Hospitality Quarterly*, 59(2), 135–159.	• This article presents a systematic review of peer-reviewed journal articles related to multigenerations within the hospitality workplace. • Methods used were very systematic, following standard procedures for systematic reviews. The authors identified 49 articles published from 2000 to 2016. • Most studies reviewed are cross-sectional. • "Certain patterns on levels of job satisfaction, commitment, turnover intentions, and [organizational citizenship behavior] among different generations were apparent. In particular, several studies showed baby boomers followed by generation X to obtain higher level of desired organizational outcomes, such as higher job satisfaction, higher commitment, higher time spent in job, and lower turnover intention, compared with generation Y. Other studies reported similar findings from the angle of generation Y, showing lower job satisfaction, lower commitment, and higher turnover compared with other generations" (p. 146).

TABLE A-1 Continued

Sources	Methods, Major Findings, Conclusions
12. Stevanin, S., Palese, A., Bressan, V., Vehviläinen-Julkunen, K., and Kvist, T. (2018). Workplace-related generational characteristics of nurses: A mixed-method systematic review. *Journal of Advanced Nursing*, 74(6), 1245–1263.	• This is a systematic review using standard methods. • Thirty-three studies met the inclusion criteria, with three main themes: (1) job attitudes, (2) emotion-related job aspects, and (3) practice- and leadership-related aspects. • "Twenty-one (63.6%) studies used a quantitative design, five (15.2%) a qualitative design, three (9.1%) a triangulated methodology with both a qualitative and quantitative design, and one (3%) a mixed method design; three (9.1%) did not report the design used" (p. 1248). • "Among the quantitative studies, only one was longitudinal... while the others were cross-sectional; the qualitative studies used primarily explorative, descriptive, and phenomenological designs" (p. 1248). • "Some intergenerational differences in workplace-related themes and subthemes emerged in the findings consistently, while others reported conflicting results" (p. 1258).
Commentaries and Methodological Validations	
13. Costanza, D.P., Darrow, J.B., Yost, A.B., and Severt, J.B. (2017). A review of analytical methods used to study generational differences: Strengths and limitations. *Work, Aging and Retirement*, 3(2), 149–165.	• Reviews and assesses, through analyses of secondary data, "the most common analytical methods that have been used in studying generational differences in social science research...group comparisons using cross-sectional data, cross-temporal meta-analysis using time-lagged panels, and cross-classified hierarchical linear modeling using time-lagged panels" (p. 149). • The purpose of the review was "to provide evidence about the extent to which the analytic methods that have been used affect the conclusions drawn about possible differences among generational groups" (p. 153). • The authors "found that each analytic method produced slightly different results, yet none was able to fully capture differences attributable to generational membership" (p. 149).

TABLE A-1 Continued

Sources	Methods, Major Findings, Conclusions
14. Parry, E. and Urwin, P. (2017). The evidence base for generational differences: Where do we go from here? *Work, Aging and Retirement*, 3(2), 140–148.	• Identifies "methodological challenges that highlight the inappropriateness of cross-sectional designs in the study of generations, as it is impossible to identify whether generation, period, or age effects are driving differences between any age groups surveyed. Second, and more fundamentally, [the authors] argue that the approach taken across most generational studies is methodologically flawed, even when more appropriate datasets are used" (p. 141). "Ultimately, this derives from a 'gap' that exists between the theoretical underpinnings claimed for this work and the empirical approaches that have been adopted in recent years" (p. 144). • The authors illustrate this gap by "using historical longitudinal data and, second, by looking for patterns within the data rather than applying generational categories a priori" (p. 141). They find that "cohort effects are predominant in early years, but then age effects dominate to demonstrate that different cohorts become less dissimilar in their 40s and beyond" (p. 145). • The authors "suggest that the patterns identified to date are simply a reflection of long-term trends in society rather than proposed differences between generational cohorts" (p. 145).
15. Rudolph, C.W., and Zacher, H. (2017). Considering generations from a lifespan developmental perspective. *Work, Aging and Retirement*, 3(2), 113–129.	• The authors "extend recent critiques of research on generations in the work context by proposing a differentiated lifespan developmental perspective" (p. 113). • The authors argue that "traditional sociological perspectives on generations are too deterministic and reductionist for understanding psychological phenomena concerning work and aging" (p. 120). They suggest "that a more contemporary model for understanding generations must be grounded in the traditions of lifespan developmental contextualism" (p. 120) and should be used to guide research with the following propositions • "Proposition 1. Historical and sociocultural contexts impact experiences and behavior at the individual level, not as shared generational effects. • Proposition 2. Developmental contextualism implies that age, period, and cohort effects are codetermined and inherently inextricable. • Proposition 3. A contextualized understanding of individual lifespan development necessitates alternative operationalizations of age, period, and cohort effects" (p. 120).

[a] Captures the difference in standard deviations between two groups (d = 0.20 = small; d = 0.50 = moderate; and d = 0.80 = large).

[b] The committee observes that "dramatically" is too strong given a d of −.13. Most would call that less than a small effect. If the confidence interval includes 0, then one cannot reject the idea that d = .00.

The studies identified through the above search process are quite international in scope, with more than 25 countries represented among samples and authors; however, most of these studies use U.S. generational categories. A range of industries, from counseling to transportation, are represented—one to three articles each except for nursing and the hospitality industry, for which there are significantly more studies (see the listing of cross-sectional studies later in this appendix). The following list identifies journals with five or more articles regarding generations and work-related outcomes:

- *Career Developmental International*
- *Employee Relations*
- *Industrial and Commercial Training*
- *Industrial and Organizational Psychology: Perspectives on Science and Practice* (special issue on this topic in 2015)
- *International Journal of Contemporary Hospitality Management*
- *International Journal of Hospitality Management*
- *Journal of Advanced Nursing*
- *Journal of Business and Psychology* (special issue on this topic in 2010)
- *Journal of Intergenerational Relationships*
- *Journal of Managerial Psychology* (special issue on this topic in 2015)
- *Journal of Nursing Administration*
- *Journal of Nursing Management*
- *Work, Aging and Retirement* (special issue on this topic in 2017)

The following are examples of topics/constructs covered in the articles:

- age and perceptions of hiring, earnings inequality, ageism in young workers, reverse ageism;
- anticipated and perceived organizational support;
- balance in work–family and work–leisure, conflict and synergy;
- communication styles, knowledge sharing;
- distress and negative social environments, mistreatment;
- employee engagement and motivation;
- generational identities;
- leadership styles and preferences;
- social contracts, psychological contracts;
- values—organizational, work, career;
- work satisfaction, burnout, turnover;
- research methodology, analysis, and theory for studying generations;

- millennials and stereotypes, archetypes, employee development and commitment, turnover factors, managers' perceptions of, leading millennials, characteristics of, health care motivations of, sense of entitlement; and
- generation Y and female leaders, employee expectations, tenure in hospitality industry, nursing, preference for place of residence, empowerment, competencies, and satisfaction.

PILOT REVIEW

The committee's pilot review was conducted on 14 articles (listed in Table A-2) selected randomly from our first set of articles (306 articles minus 57 of those identified as irrelevant to this study) as follows:

Round 1: Every 7th article out of the 249 articles, sorted alphabetically by first author, dated 2000 or later, generating 35 articles for Round 2. Round 2: Every 3rd article out of the subset of 35, sorted by date from oldest to newest, yielding 11 articles. In this draw, there was only 1 article with a research design that aimed to separate cohort effects from age or period effects (i.e., other than a cross-sectional or qualitative design). We added 3 articles to oversample articles using other statistical methods. The final draw resulted in 14 articles.

Based on our understanding from previous reviews, as well as our own knowledge of studies in this area, we believed the sample of articles generated for our pilot review allowed us to appreciate the different research designs used in this literature and a mix of conclusions with regard to generational differences. Cross-sectional designs were prominent (eight studies). We were initially surprised that eight studies were conducted outside of the United States but have come to appreciate that a large percentage of the generational research is conducted in other countries. Many of these international studies use the U.S. generational labels (e.g., baby boomers and millennials) to categorize their groups. These articles acknowledge the limits of these labels, often with a note recognizing that people in their countries would have had different experiences at different times.

For our pilot review, we developed a coding scheme. Two members of the committee and two National Academies staff manually coded the following characteristics of each article: (1) author(s) and publication year, title, and source; (2) country/countries in which the study was conducted; (3) data source (primary versus secondary data) and type (quantitative or qualitative); (4) sampling strategy (nonprobability, probability); (5) sample type (e.g., student or working adult) and associated industry if relevant;

TABLE A-2 Articles in the Committee's Pilot Review

Reference	Study Type
Antonczyk, D., DeLeire, T., and Fitzenberger, B. (2018). Polarization and rising wage inequality: Comparing the U.S. and Germany. *Econometrics*, 6(2), 1–33.	The study examines wage inequality in the United States and Germany using nationally representative survey data from 1979 to 2004 (probability samples) and an approach developed by MaCurdy and Mroz (1995) to separate age, time, and cohort effects.
Cennamo, L., and D. Gardner (2008). Generational differences in work values, outcomes and person-organisation values fit. *Journal of Managerial Psychology*, 23(8), 891–906.	The study is cross-sectional and based on self-report data, limiting the generalizability of findings. A total of 504 Auckland employees representing a range of industries completed an online questionnaire.
Danigelis, N. L., Hardy, M., and Cutler, S. (2007). Population aging, intracohort aging, and sociopolitical attitudes. *American Sociological Review*, 2(5), 812–830.	The article examines attitude change in the U.S. population using data from the General Social Survey, 1972–2004, but the focus is not work-related or generational.
Heritage, B., Breen, L., and Roberts, L. D. (2016). In-groups, out-groups, and their contrasting perceptions of values among generational cohorts of Australians. *Australian Psychologist*, 51(3), 246–255.	The study examines and compares self-ratings and out-group perceptions of the importance of the four overarching clusters of values in Schwartz's circumplex model by generation. A convenience sample of 157 participants completed an online survey of self-rated values and perceptions of another generation's values.
Krajcsák, Z., Jonás, T., and Henrietta, F. (2014). An analysis of commitment factors depending on generation and part-time working in selected groups of employees in Hungary. *Argumenta Oeconomica*, 33(2), 115–144.	The study is cross-sectional, based on data from 661 respondents to a questionnaire designed to analyze factors related to employee commitment.
Lyons, S.T., Duxbury, L., and Higgins, C. (2007). An empirical assessment of generational differences in basic human values. *Psychological Reports*, 101(2), 339–352.	This cross-sectional study assesses generational differences in human values as measured by the Schwartz Value Survey among a combined sample of Canadian knowledge workers and undergraduate business students (N = 1,194).
Raineri, N., Paillé, P., and Morin, Denis. (2012). Organizational citizenship behavior: An intergenerational study. *Revue Internationale de Psychologie Sociale*, 25(3–4), 147–177.	The authors use social exchange theory to investigate whether membership in the baby boomer versus generation X group influences the relationships of organization- and colleague-directed support and commitment with organizational citizenship behavior, and uses structural equation modeling to analyze data from voluntary survey responses (N = 943).

TABLE A-2 Continued

Reference	Study Type
Singh, U., and Weimar, D. (2017). Empowerment among generations. *German Journal of Human Resource Management*, 31(4), 307–328.	This cross-sectional study investigated differences in people's attitudes toward empowerment by generation and other demographic variables using survey data from a convenience sample (N = 492).
Soni, S., Upadhyaya, M., and Kautish, P. (2011). Generational differences in work commitment of software professionals: Myth or reality? *Abhigyan*, 28(4), 30–42.	This cross-sectional study examined generational differences for five types of work commitment. A total of 250 respondents working in software industries were administered a questionnaire.
Takase, M., Oba, K., and Yamashita, N. (2009). Generational differences in factors influencing job turnover among Japanese nurses: An exploratory comparative design. *International Journal of Nursing Studies*, 46(7), 957–967.	The purpose of the study was to identify specific work-related needs and values of nurses in three generations. The study was conducted in three public hospitals in Japan. A convenience sample of 315 registered nurses participated. A survey was used to collect quantitative and qualitative data.
Trzesniewski, K.H., and Donnellan, M.B. (2010). Rethinking "generation me": A study of cohort effects from 1976–2006. *Perspectives on Psychological Science*, 5(1), 58–75.	This is a study of cohort effects using large samples of U.S. high school seniors from 1976 to 2006 from the Monitoring the Future program (total N = 477,380). The goal of the study was to test the strength of cohort effects on 31 psychological constructs.
Twenge, J M., Konrath, S., Foster, J.D., Keith Campbell, W., and Bushman, B.J. (2008). Egos inflating over time: A cross temporal meta-analysis of the Narcissistic Personality Inventory. *Journal of Personality*, 76(4), 875–902.	This was a cross-temporal meta-analysis with time-lagged data from 85 samples of American college students who completed the Narcissistic Personality Inventory between 1979 and 2006 (N = 16,475).
Baker Rosa, N.M., and Hastings, S.O. (2018). Managing millennials: Looking beyond generational stereotypes. *Journal of Organizational Change Management*, 31(4), 920–930.	The purpose of this qualitative study was to examine managers' perceptions of millennial employees in organizations. In total, 25 interviews were conducted with managers in the hospitality industry.
Milner, S., Demilly, H., and Pochic, S. (2019). Bargained equality: The strengths and weaknesses of workplace gender equality agreements and plans in France. *British Journal of Industrial Relations*, 57(2), 75–301.	This study evaluated a sample of 146 workplace agreements and plans on gender equality submitted in 2014–2015, in 10 sectors, and involved in-depth interviews in 20 companies. The study examined "generational effects" in terms of process change and not differences among workers.

(6) respondent demographic characteristics as reported (e.g., age group-ings, men/women, and race/ethnicity); and (7) study design. Two articles (Danigelis et al., 2007; Milner et al., 2019) were determined to be irrelevant for our task.

There were no disagreements among the four reviewers on the nature of the publications, just opportunities to clarify what we were observing among the reviewers, as well as with the larger committee. Initially, we attempted to capture authors' definitions of generation, proposed anteced-ents to hypothesized differences, and theoretical approach, but this effort proved to be unsatisfactory. None of the authors provides an antecedent, that is, defining events or a set of experiences that would have influenced generations. With some exceptions, they generally point to earlier sociologi-cal theories on generational change and assume that generational influences would be responsible for observed differences. All authors use birth cohorts to define generations, and most use popular generational terminology (e.g., baby boomers, millennials) to label groups of workers.

Most of the quantitative studies are cross-sectional, having used convenience samples to collect primary data from self-reports on questionnaires, and in some cases through interviews. Several of the studies utilized snowball sampling (e.g., Krajcsak et al., 2014). Most of these studies examined work-related attitudes/values, with a noticeable focus on "commitment" (e.g., Raineri et al., 2012). Many of the pilot articles break findings down by men and women. Only one article notes that most of the sample was Caucasian; otherwise, the race/ethnicity of the samples is not considered. Notably, a few studies measure educational attainment, employment status, and/or life stage (e.g., have children). The findings, whether age or generation effects, across this pilot sample are mixed, with some authors reporting that their findings indicate differences among generation groups and others finding no differences on the measured values/attitudes. One study (Trzesniewski and Donnellan, 2010) used secondary, nationally representative data and cohort analysis to test the plausibility of previously reported cohort effects on psychological constructs. These authors found little evidence for significant distinctions among generations.

SAMPLE OF GENERATIONAL LITERATURE

The following list of references is intended to illustrate the types of empirical studies the committee found in assembling the literature related to generational attitudes and behaviors in the workforce. The list is organized by research design and then alphabetically by first author. For a full list of articles identified for this report, visit https://nas.edu/workforcegenerations.

Multilevel Models Applied to Nested Datasets (APC Models)

Multilevel models are a family of statistical tools appropriate for studying databases in which some observations are nested within others, such as when multiple individuals provide data in different years, as in the case of cross-sectional studies repeated across multiple years. Statistically speaking, individual responses then are nested within each year. Likewise, nesting can occur in longitudinal studies when the same people are observed repeatedly over time. In this case, observations on different occasions are nested within people.

Donnelly, K., Twenge, J., Clark, M., Shaikh, S., Beiler-May, A., and Carter, N. (2016). Attitudes toward women's work and family roles in the United States, 1976–2013. *Psychology of Women Quarterly*, 40(1), 41–54.

Jürges, H. (2003). Age, cohort, and the slump in job satisfaction among West German workers. *Labour*, 17(4), 489–518.

Kalleberg, A. L., and Marsden, P.V. (2019). Work values in the United States: Age, period, and generational differences. *Annals of the American Academy*, 682(1), 43–59.

Koning, P., and Raterink, M. (2013). Re-employment rates of older unemployed workers: Decomposing the effect of birth cohorts and policy changes. *Economist (Netherlands)* 161(3), 331–348.

Kowske, B., Rasch, R., and Wiley, J. (2010). Millennials' (lack of) attitude problem: An empirical examination of generational effects on work attitudes. *Journal of Business and Psychology*, 25, 265–279. [Data from employees in the United States.]

Cross-Temporal Meta-Analyses

Cross-temporal meta-analyses entail extracting descriptive statistics (often measures of central tendency, such as sample means) from studies conducted at different points in time. These descriptive statistics are combined using meta-analytic techniques and usually weighted for precision by the number of observations available for each time point. The objective is to test whether aggregated estimates vary because of when the data were collected.

Campbell, S.M., Twenge, J.M., and Campbell, W.K. (2017). Fuzzy but useful constructs: Making sense of the differences between generations. *Work, Aging and Retirement*, 3(2), 130–139.

Twenge, J.M., and Campbell, W.K. (2001). Age and birth cohort differences in self-esteem: A cross-temporal meta-analysis. *Personality and Social Psychology Review*, 5, 321–344.

Twenge, J.M., and Campbell, S.M. (2008). Generational differences in psychological traits and their impact on the workplace. *Journal of Managerial Psychology*, 23(8), 862–877.

Twenge, J.M., Freeman, E.C., and Campbell W.K. (2012). Generational differences in young adults' life goals, concern for others, and civic orientation, 1966–2009. *Journal of Personality and Social Psychology*, 102(5), 1045–1062.

Other Studies Comparing Samples over Time[1]

Hansen, J.I.C., and Leuty, M.E. (2012). Work values across generations. *Journal of Career Assessment*, 20(1), 34–52.

Krahn, H.J., and Galambos, N.L. (2014). Work values and beliefs of "Generation X" and "Generation Y." *Journal of Youth Studies*, 17(1), 92–112.

Leuty, M.E., and Hansen, J.I.C. (2014). Teasing apart the relations between age, birth cohort, and vocational interests. *Journal of Counseling Psychology*, 61(2), 289–298.

Lippmann, S. (2008). Rethinking risk in the new economy: Age and cohort effects on unemployment and reemployment. *Human Relations*, 61, 1259–1292.

Smola, K.W., and Sutton, C.D. (2002). Generational differences: Revisiting generational work values for the new millennium. *Journal of Organizational Behavior*, 23(SPEC. ISS.), 363–382.

Teclaw, R., Osatuke, K., Fishman, J., Moore S.C., Dyrenforth, S. (2014). Employee age and tenure within organizations: Relationship to workplace satisfaction and workplace climate perceptions. *Health Care Manager*, 33(1), 4–19.

Trzesniewski, K.H., and Donnellan, M.B. (2010). Rethinking "generation me": A study of cohort effects from 1976–2006. *Perspectives on Psychological Science* 5(1), 58–75.

Twenge, J.M., Campbell, S.M., Hoffman, B.J., and Lance, C.E. (2010). Generational differences in work values: Leisure and extrinsic values increasing, social and intrinsic values decreasing. *Journal of Management*, 36(5), 1117–1142.

Cross-Sectional Designs

Cross-sectional research designs compare groups of people of different ages using an instrument (e.g., a survey) administered to a single sample at a single point in time. The following list includes 46 of the more than 300 cross-sectional studies the committee identified—those cited in the main text of this report, notably the studies for particular types of jobs.

Nursing

Andrews, D.R. (2013). Expectations of millennial nurse graduates transitioning into practice. *Nursing Administration Quarterly*, 37(2), 152–159. doi: https://doi.org/10.1097/NAQ.0b013e3182869d9f.

Anthony, M.K., Tullai-McGuinness, S., Capone, L., and Farag, A. (2008). Decision making, autonomy, and control over practice: Are there variations across generational cohorts? *Journal of Nursing Administration*, 38(5), 211.

Carver, L., and Candela, L. (2008). Attaining organizational commitment across different generations of nurses. *Journal of Nursing Management*, 16(8), 984–991. doi: https://doi.org/10.1111/j.1365-2834.2008.00911.x.

Chung, S.M., and Fitzsimons, V. (2013). Knowing generation Y: A new generation of nurses in practice. *British Journal of Nursing*, 22(20), 1173–1179.

---

[1] These studies vary in their approach to measuring variance in work values in groups of people over time. One is longitudinal in that the authors collected data from the same people at different points in time. Others analyze data from different groups of similar participants (usually by age) collected at different points in time on the same constructs. Three of these use data from nationally representative surveys (Current Population Survey, General Social Survey, and Monitoring the Future); others use survey data from smaller samples.

Clendon, J., and Walker, L. (2012). "Being young": A qualitative study of younger nurses' experiences in the workplace. *International Nursing Review,* 59(4), 555–561. doi: https://doi.org/10.1111/j.1466-7657.2012.01005.x.

Crowther, A., and Kemp, M. (2009). Generational attitudes of rural mental health nurses. *Australian Journal of Rural Health,* 17(2), 97–101. doi: https://doi.org/10.1111/j.1440-1584.2009.01044.x.

Farag, A.A., Tullai-Mcguinness, S., and Anthony, M K. (2009). Nurses' perception of their manager's leadership style and unit climate: Are there generational differences? *Journal of Nursing Management,* 17(1), 26–34. doi: https://doi.org/10.1111/j.1365-2834.2008.00964.x.

Hamlin, L., and Gillespie, B.M. (2011). Beam me up, Scotty, but not just yet: Understanding generational diversity in the perioperative milieu. *Journal of Perioperative Nursing,* 24(4), 36–43.

Hendricks, J.M., and Cope, V.C. (2013). Generational diversity: What nurse managers need to know. *Journal of Advanced Nursing,* 69(3), 717–725. doi: https://doi.org/10.1111/j.1365-2648.2012.06079.x.

Hu, J., Herrick, C., and Hodgin, K.A. (2004). Managing the multigenerational nursing team. *Health Care Manager,* 23(4), 334–340. doi: https://doi.org/10.1097/00126450-200410000-00008.

Keepnews, D.M., Brewer, C.S., Kovner, C.T., and Shin, J.H. (2010). Generational differences among newly licensed registered nurses. *Nursing Outlook,* 58(3), 155–163. doi: https://doi.org/10.1016/j.outlook.2009.11.001.

Lavoie-Tremblay, M., Trépanier, S.G., Fernet, C., and Bonneville-Roussy, A. (2014). Testing and extending the triple match principle in the nursing profession: A generational perspective on job demands, job resources and strain at work. *Journal of Advanced Nursing,* 70(2), 310–322. doi: https://doi.org/10.1111/jan.12188.

Leiter, M.P., Jackson, N.J., and Shaughnessy, K. (2009). Contrasting burnout, turnover intention, control, value congruence and knowledge sharing between baby boomers and generation X. *Journal of Nursing Management,* 17(1), 100–109. doi: http://dx.doi.org/10.1111/j.1365-2834.2008.00884.x.

Leiter, M.P., Price, S.L., and Spence Laschinger, H.K. (2010). Generational differences in distress, attitudes and incivility among nurses. *Journal of Nursing Management,* 18(8), 970–980. doi: https://doi.org/10.1111/j.1365-2834.2010.01168.x.

LeVasseur, S.A., Wang, C.Y., Mathews, B., and Boland, M. (2009). Generational differences in registered nurse turnover. *Policy, Politics, and Nursing Practice,* 10(3), 212–223. doi: https://doi.org/10.1177/1527154409356477.

Nelson, S.A. (2012). Affective commitment of generational cohorts of Brazilian nurses. *International Journal of Manpower,* 33(7), 804–821. doi: https://doi.org/10.1108/01437721211268339.

Santos, S.R., and Cox, K. (2000). Workplace adjustment and intergenerational differences between matures, boomers, and xers. *Nursing Economic$,* 18(1), 7–13.

Shacklock, K., and Brunetto, Y. (2012). The intention to continue nursing: Work variables affecting three nurse generations in Australia. *Journal of Advanced Nursing,* 68(1), 36–46. doi: http://dx.doi.org/10.1111/j.1365-2648.2011.05709.x.

Sparks, A.M. (2012). Psychological empowerment and job satisfaction between baby boomer and generation X nurses. *Journal of Nursing Management,* 20(4), 451–460. doi: https://doi.org/10.1111/j.1365-2834.2011.01282.x.

Takase, M., Oba, K., and Yamashita, N. (2009). Generational differences in factors influencing job turnover among Japanese nurses: An exploratory comparative design. *International Journal of Nursing Studies,* 46(7), 957–967. doi: https://doi.org/10.1016/j.ijnurstu.2007.10.013.

Thompson, J.A. (2007). Why work in perioperative nursing? Baby boomers and generation Xers tell all. *AORN Journal:The Official Voice of Perioperative Nursing*, 86(4), 564–587. doi: http://dx.doi.org/10.1016/j.aorn.2007.03.010.

Tourangeau, A.E., Thomson, H., Cummings, G., and Cranley, L.A. (2013). Generation-specific incentives and disincentives for nurses to remain employed in acute care hospitals. *Journal of Nursing Management*, 21(3), 473–482. doi: http://dx.doi.org/10.1111/j.1365-2834.2012.01424.x.

Tourangeau, A.E., Wong, M., Saari, M., and Patterson, E. (2015). Generation-specific incentives and disincentives for nurse faculty to remain employed. *Journal of Advanced Nursing*, 71(5), 1019–1031. doi: http://dx.doi.org/10.1111/jan.12582.

Wakim, N. (2014). Occupational stressors, stress perception levels, and coping styles of medical surgical RNs: A generational perspective. *Journal of Nursing Administration*, 44(12), 632–639. doi: https://doi.org/10.1097/NNA.0000000000000140.

Warshawski, S., Barnoy, S., and Kagan, I. (2017). Professional, generational, and gender differences in perception of organisational values among Israeli physicians and nurses: Implications for retention. *Journal of Interprofessional Care*, 31(6), 696–704. doi: http://dx.doi.org/10.1080/13561820.2017.1355780.

Wilson, B., Squires, M., Widger, K., Cranley, L., and Tourangeau, A. (2008). Job satisfaction among a multigenerational nursing workforce. *Journal of Nursing Management*, 16(6), 716–723. doi: https://doi.org/10.1111/j.1365-2834.2008.00874.x.

Hospitality

Arendt, S.W., Roberts, K.R., Strohbehn, C., Arroyo, P.P., Ellis, J., and Meyer, J. (2014). Motivating foodservice employees to follow safe food handling practices: Perspectives from a multigenerational workforce. *Journal of Human Resources in Hospitality and Tourism*, 13(4), 323–349. doi: https://doi.org/10.1080/15332845.2014.888505.

Barron, P., Leask, A., and Fyall, A. (2014). Engaging the multi-generational workforce in tourism and hospitality. *Tourism Review*, 69(4), 245–263. doi: https://doi.org/10.1108/TR-04-2014-0017.

Bednarska, M.A. (2016). Complementary person-environment fit as a predictor of job pursuit intentions in the service industry. *Contemporary Economics*, 10(1), 27–38. doi: https://doi.org/10.5709/ce.1897-9254.196.

Chen, P.J., and Choi, Y. (2008). Generational differences in work values: A study of hospitality management. *International Journal of Contemporary Hospitality Management*, 20(6), 595–615. doi: https://doi.org/10.1108/09596110810892182.

Choi, Y.G., Kwon, J., and Kim, W. (2013). Effects of attitudes vs experience of workplace fun on employee behaviors: Focused on Generation Y in the hospitality industry. *International Journal of Contemporary Hospitality Management*, 25(3), 410–427. doi: https://doi.org/10.1108/09596111311311044.

Goh, E., and Lee, C. (2018). A workforce to be reckoned with: The emerging pivotal generation Z hospitality workforce. *International Journal of Hospitality Management*, 73, 20–28. doi: https://doi.org/10.1016/j.ijhm.2018.01.016.

Gursoy, D., Chi, C.G.Q., and Karadag, E. (2013). Generational differences in work values and attitudes among frontline and service contact employees. *International Journal of Hospitality Management*, 32(1), 40–48. doi: https://doi.org/10.1016/j.ijhm.2012.04.002.

Kim, M., Knutson, B.J., and Choi, L. (2016). The effects of employee voice and delight on job satisfaction and behaviors: Comparison between employee generations. *Journal of Hospitality Marketing and Management*, 25(5), 563–588. doi: https://doi.org/10.1080/19368623.2015.1067665.

King, C., Murillo, E., and Lee, H. (2017). The effects of generational work values on employee brand attitude and behavior: A multi-group analysis. *International Journal of Hospitality Management, 66,* 92–105. doi: http://dx.doi.org/10.1016/j.ijhm.2017.07.006.

Kong, H., Sun, N., and Yan, Q. (2016). New generation, psychological empowerment: Can empowerment lead to career competencies and career satisfaction? *International Journal of Contemporary Hospitality Management, 28*(11), 2553–2569. doi: https://doi.org/10.1108/IJCHM-05-2014-0222.

Kong, H., Wang, S., and Fu, X. (2015). Meeting career expectation: Can it enhance job satisfaction of generation Y? *International Journal of Contemporary Hospitality Management, 27*(1), 147–168. doi: https://doi.org/10.1108/IJCHM-08-2013-0353.

Lu, A.C.C., and Gursoy, D. (2016). Impact of job burnout on satisfaction and turnover intention: Do generational differences matter? *Journal of Hospitality and Tourism Research, 40*(2), 210–235. doi: https://doi.org/10.1177/1096348013495696.

Lub, X.D., Blomme, R.J., and Matthijs Bal, P. (2011). Psychological contract and organizational citizenship behavior: A new deal for new generations? *Advances in Hospitality and Leisure, 7,* 109–130.

Lub, X., Bijvank, M.N., Bal, P.M., Blomme, R., and Schalk, R. (2012). Different or alike?: Exploring the psychological contract and commitment of different generations of hospitality workers. *International Journal of Contemporary Hospitality Management, 24*(4), 553–573. doi: https://doi.org/10.1108/09596111211226824.

Lub, X., Bal, P.M., Blomme, R.J., and Schalk, R. (2016). One job, one deal...or not: Do generations respond differently to psychological contract fulfillment? *The International Journal of Human Resource Management 27*(6), 653–680. doi: http://dx.doi.org/10.10 80/09585192.2015.1035304.

Maier, T.A. (2011). Hospitality leadership implications: Multigenerational perceptions of dissatisfaction and intent to leave. *Journal of Human Resources in Hospitality and Tourism, 10*(4), 354–371. doi: https://doi.org/10.1080/15332845.2011.588503.

Park, J., and Gursoy, D. (2012). Generation effects on work engagement among U.S. hotel employees. *International Journal of Hospitality Management, 31*(4), 1195–1202. doi: https://doi.org/10.1016/j.ijhm.2012.02.007.

Supanti, D., and Butcher, K. (2019). Is corporate social responsibility (CSR) participation the pathway to foster meaningful work and helping behavior for millennials? *International Journal of Hospitality Management, 77,* 8–18. doi: https://doi.org/10.1016/j.ijhm.2018.06.001.

Tsaur, S.H., and Yen, C.H. (2018). Work–leisure conflict and its consequences: Do generational differences matter? *Tourism Management, 69,* 121–131. doi: https://doi.org/10.1016/j.tourman.2018.05.011.

Zopiatis, A., Krambia-Kapardis, M., and Varnavas, A. (2012). Y-ers, X-ers and boomers: Investigating the multigenerational (mis)perceptions in the hospitality workplace. *Tourism and Hospitality Research, 12*(2), 101–121. doi: https://doi.org/10.1177/1467358412466668.

Qualitative Studies

The 13 articles marked with an asterisk are focused more on understanding one generation and do not compare generations.

Abdul Malek, M.M., and A.R. Jaguli. (2018). Generational differences in workplace communication: Perspectives of female leaders and their direct reports in Malaysia. *Journal of Asian Pacific Communication, 28*(1), 129–150.

Andrews, D.R. (2013). Expectations of millennial nurse graduates transitioning into practice. *Nursing Administration Quarterly*, 37(2), 152–159.*

Baker Rosa, N.M., and S O. Hastings (2018). Managing millennials: Looking beyond generational stereotypes. *Journal of Organizational Change Management*, 31(4), 920–930.*

Bone, Z., and K. Tilbrook (2015). Women as bosses: A snapshot from a generational perspective. *International Journal of Organizational Diversity*, 15(3), 13–24.

Boyd, D. (2010). Ethical determinants for generations X and Y. *Journal of Business Ethics*, 93(3), 465–469.

Brown, E., Thomas N., and Bosselman, R. (2015). Are they leaving or staying: A qualitative analysis of turnover issues for Generation Y hospitality employees with a hospitality education. *International Journal of Hospitality Management*, 46, 130–137.*

Chillakuri, B., and Mogili, R. (2018). Managing millennials in the digital era: Building a sustainable culture. *Human Resource Management International Digest*, 26(3), 7–10.*

Clarke, M. (2015). Dual careers: The new norm for Gen Y professionals? *Career Development International*, 20(6), 562–582.*

Clendon, J., and L. Walker (2012). Being young: A qualitative study of younger nurses' experiences in the workplace. *International Nursing Review*, 59(4), 555–561.*

Feyerherm, A., and Vick, Y.H. (2005). Generation X women in high technology: Overcoming gender and generational challenges to succeed in the corporate environment. *The Career Development International*, 10(3), 216–227.*

Foster, K. (2013). Generation and discourse in working life stories. *British Journal of Sociology*, 64(2), 195–215.

Gale, D. (2013). Career resumption for educated baby boomer mothers: An exploratory study. *Journal of Intergenerational Relationships*, 11(3), 304–319.*

Goh, E., and Lee, C. (2018). A workforce to be reckoned with: The emerging pivotal generation Z hospitality workforce. *International Journal of Hospitality Management* 73, 20–28.*

Gursoy, D., Maier, T., and Chi, C. (2008). Generational differences: An examination of work values and generational gaps in the hospitality workforce. *International Journal of Hospitality Management*, 27(3), 448–458.

Haapala, I., Tervo, L., and Biggs, S. (2015). Using generational intelligence to examine community care work between younger and older adults. *Journal of Social Work Practice*, 29(4), 457–473.

James, L. (2009). Generational differences in women's attitudes towards paid employment in a British city: The role of habitus. *Gender, Place and Culture*, 16(3), 313–328.

Kultalahti, S., and Viitala, R. (2015). Generation Y—Challenging clients for HRM? *Journal of Managerial Psychology*, 30(1), 101–114.*

Lee, S. (2014). Korean mature women students' various subjectivities in relation to their motivation for higher education: Generational differences amongst women. *International Journal of Lifelong Education*, 33(6), 791–810.

Lyons, S.T., and Schweitzer, L. (2017). A qualitative exploration of generational identity: Making sense of young and old in the context of today's workplace. *Work, Aging and Retirement*, 3(2), 209–224.

Matthews, M., Seguin, M., Chowdhury, N., and Card, R. (2012). Generational differences in factors influencing physicians to choose a work location. *Rural and Remote Health*, 12(1).

Patel, J., Tinker, A., and Corna, L. (2018). Younger workers' attitudes and perceptions towards older colleagues. *Working with Older People*, 22(3), 129–138.

Price, S., McGillis Hall, L., Murphy, G., and Pierce, B. (2018). Evolving career choice narratives of new graduate nurses. *Nurse Education in Practice*, 28, 86–91.*

Pritchard, K., and Whiting, R. (2014). Baby boomers and the lost generation: On the discursive construction of generations at work. *Organization Studies*, 35(11), 1605–1626.

Sanders, M.J., and McCready, J. (2009). A qualitative study of two older workers' adaptation to physically demanding work. *Work*, 32(2), 111–122.*

Singh, V. (2013). Exploring the concept of work across generations. *Journal of Intergenerational Relationships*, 11(3), 272–285.

Stone-Johnson, C. (2014). Not cut out to be an administrator: Generations, change, and the career transition from teacher to principal. *Education and Urban Society*, 46(5), 606–625.*

Urick, M.J., Hollensbe, E.C., Masterson, S.S., and Lyons, S.T., (2017). Understanding and managing intergenerational conflict: An examination of influences and strategies. *Work, Aging and Retirement*, 3(2), 166–185.

Whitmer, M., Hurst, S., and Prins, M. (2009). Intergenerational views of hardiness in critical care nurses. *Dimensions of Critical Care Nursing*, 28(5), 214–220.

Williams, G. (in press). Management Millennialism: Designing the new generation of employee. *Work, Employment and Society*. doi: https://doi.org/10.1177/0950017019836891.

Mixed Methods

Gardiner, S., Grace, D., and King, C. (2013). Challenging the use of generational segmentation through understanding self-identity. *Marketing Intelligence and Planning*, 31(6), 639–653.

Gordon, P.A. (2017). Exploring generational cohort work satisfaction in hospital nurses. *Leadership in Health Services*, 30(3), 233–248.

Kwiek, M. (2017). A generational divide in the academic profession: A mixed quantitative and qualitative approach to the Polish case. *European Educational Research Journal*, 16(5), 645–669.

Van Rossem, A.H.D. (2019). Generations as social categories: An exploratory cognitive study of generational identity and generational stereotypes in a multigenerational workforce. *Journal of Organizational Behavior*, 40(4), 434–455.

Weeks, K.P., and Schaffert, C. (2019). Generational differences in definitions of meaningful work: A mixed methods study. *Journal of Business Ethics*, 156(4), 1045–1061.

Zimmerer, T.E. (2013). *Generational Perceptions of Servant Leadership: A Mixed Methods Study*. (Doctoral dissertation, Capella University).

Appendix B

Biographical Sketches of Committee Members and Staff

Nancy T. Tippins (*Chair*) is a principal at The Nancy T. Tippins Group, LLC, Greenville, South Carolina. As an industrial and organizational psychologist, she has worked as both an internal and external consultant to a variety of industries. Her work encompasses the study of employment practices, including executive coaching, job analysis, competency development, selection, training, manager and executive assessment, employee and management development, succession planning, compensation, complaint procedures, and other policies and procedures related to equal employment opportunity. For the Society for Industrial and Organizational Psychology, she has served as president, on the ad hoc committee for the revision of the *Principles for the Validation and Use of Personnel Selection Procedures*, and as cochair of the committee for the most recent revision of the *Principles*. She was one of the U.S. representatives on the International Standardization Organization committee to establish international testing standards. She has a B.A. in history from Agnes Scott College, an M.Ed. in counseling and psychological services from Georgia State University, and an M.S. and a Ph.D. in industrial and organizational psychology from the Georgia Institute of Technology.

Eric M. Anderman is professor and former chair in the Department of Educational Studies at The Ohio State University, with appointments in educational psychology and the quantitative research, evaluation, and measurement program. Previously, he held faculty and administrator positions at the University of Kentucky. Although much of his research has focused on student learning and motivation, he is also interested in how learning progresses in work contexts. While serving as department chair, he administered 13 graduate programs, including a highly ranked online degree program. He has also participated in and chaired a coalition of psychologists working to translate psychological science and solve problems associated with schools and education for the American Psychological Association. He is the editor of the journal *Theory into Practice* and is a fellow of the American Psychological Association and of the American Educational Research Association. He has a Ph.D. in educational psychology from the University of Michigan.

John Baugh is the Margaret Bush Wilson professor in arts and sciences at Washington University in St. Louis, where he holds academic appointments in psychological and brain sciences, anthropology, linguistics, education, English, African and African American Studies, American culture studies, philosophy-neuroscience-psychology, and urban studies. His previous academic appointments were at Swarthmore College, the University of Texas at Austin, and Stanford University, where he is professor emeritus of education and linguistics. His research evaluates the social stratification of linguistic diversity in advanced industrial societies with relevance to matters of policy in education, medicine, and law. His work also focuses on advancing studies of linguistic profiling and various forms of linguistic discrimination. He is a fellow of the Linguistic Society of America. He received a B.A. in speech and communication from Temple University and an M.A. and a Ph.D. in linguistics from the University of Pennsylvania.

Margaret E. Beier is a professor of psychological sciences at Rice University. Her research interests lie at the intersection of aging, lifelong learning and development, and work. In particular, she examines the motivation and ability determinants of behavior at different stages of the work lifespan, from school to work and work to retirement. She also examines the effectiveness of various educational interventions and individual factors on learning, performance, and noncognitive outcomes, such as self-efficacy, self-concept, and interests. She is a member of the American Educational Research Association, and she is a fellow of the Association for Psychological Science and of the Society for Industrial and Organizational Psychologists (Division 14) of the American Psychological Association. She received a B.A. from Colby College in Waterville, Maine, and an M.S. and a Ph.D. in psychology from the Georgia Institute of Technology.

Dana H. Born is a lecturer in public policy at Harvard Kennedy School of Government. She is also codirector of the Center for Public Leadership and faculty chair of the Senior Executive Fellows Program. She is a retired brigadier general of the U.S. Air Force, and she served two terms as the dean of the faculty and department chair for the Behavioral Sciences and Leadership Department at the U.S. Air Force Academy. During her military career, she also served as an exchange officer with the Royal Australian Air Force; aide and speech writer to the Secretary of the Air Force; squadron commander at Bolling Air Force Base, Washington, DC; and in Afghanistan in support of Operation Enduring Freedom. Her experiences and training form the basis of her focus today on change and risk management, organizational behavior and ethics, character-based leadership development, human and social capacity and performance, strategic alignment, inclusive excellence, diversity, decision-making and gender related issues. She has a B.S. in behavioral sciences from the U.S. Air Force Academy, an M.S. in experimental psychology from Trinity University, an M.A. in research psychology from the University of Melbourne, and a Ph.D. in industrial and organizational psychology from Penn State University.

Chandra Childers is a study director at the Institute for Women's Policy Research, Washington, DC, an organization that advances women's status through research, policy analysis, and public education. Using both qualitative and quantitative methods, her research uses an intersectional lens to examine women's and men's employment, earnings, and job quality; the effects of technology (automation/artificial intelligence/digitalization) on current and future labor market experiences; and issues and concerns for women and girls of color. The work includes a study of unemployment during and after the Great Recession for millennial women and the effect of automation on the future of work for older women. Previously, at the University of Washington, she taught a range of courses, including the research practicum, and provided research support for projects that included employment discrimination cases. She has a M.S. in human development from Texas Tech University and a Ph.D. in sociology from the University of Washington.

Brent Donnellan is chair and a professor of the Department of Psychology at Michigan State University. Previously, he was a professor of psychology at Texas A&M University. His research investigates questions at the intersections of developmental psychology, personality psychology, and psychological assessment. He also works on methodological reform in psychological science, and he has published several papers about the size and strength of the evidence for generational shifts in individual characteristics, including personality, self-esteem, and other attitudes. He is currently a co-lead on a study investigating how experiences in the workplace are associated with personality trait development. He is the senior editor for the personality section for *Collabra: Psychology*. He has a Ph.D. in human development from the University of California, Davis.

Armando X. Estrada is an assistant professor in the Department of Policy, Organizational, and Leadership Studies at Temple University. His research centers on the assessment of job attitudes and behaviors, training development and evaluation, and strategic planning and assessment. He works on factors influencing diversity, inclusion, and engagement, with particular focus in two areas: women and minorities in civilian and military organizations and factors influencing leadership, teamwork, and performance, especially on cohesion, readiness, resilience, and effectiveness of collectives in civilian and military organizations. He previously served as a program manager and senior research psychologist with the Foundational Science Research Unit of the U.S. Army Research Institute for the Behavioral and Social Sciences. Earlier, he held various academic positions, including professorships at Washington State University, the U.S. Naval Postgraduate School, the Industrial College of the Armed Forces, and National Defense University. He has a B.S. and an M.S. in psychology from the California State University at Los Angeles and a Ph.D. in industrial and organizational psychology from the University of Texas at El Paso.

Brian Hoffman is a professor in the Department of Psychology and the chair of the industrial-organizational psychology program at the University of Georgia. His research and publications cover the changing nature of work and workers, the assessment and prediction of effective leadership, and the application of management principles to sports settings. His primary research interest revolves around the measurement and prediction of human performance, with a specific emphasis on evaluating the skills and behaviors associated with effective leadership. His work focuses in particular on measuring visionary leadership, managerial skills, and altruism at work. He is a fellow of the Society for Industrial and Organizational Psychology and currently serves as an associate editor of the *Journal of Management*. He has a Ph.D. in industrial and organizational psychology from the University of Tennessee.

Arne L. Kalleberg is the Kenan Distinguished professor of sociology, chair of the Curriculum in Global Studies, and adjunct professor of public policy and of management at the University of North Carolina at Chapel Hill. He also serves as distinguished research fellow in the Center for Strategy and Leadership at the Foundation for Research in Economics and Business Administration in Bergen, Norway. He has published extensively in the areas of the sociology of work, occupations and organizations, economy and society, and social stratification and inequality. He is editor-in-chief of *Social Forces*, an international journal of social research. He is an elected fellow of the American Association for the Advancement of Science, the Labor and Employment Relations Association, the Association for Psychological Science, and The Royal Norwegian Society of Sciences and Letters. He has a Ph.D. in sociology from the University of Wisconsin at Madison.

Ruth Kanfer is a professor of psychology in the School of Psychology at the Georgia Institute of Technology. Her research focuses on the influence of motivation, personality, and emotion in workplace behavior, job performance, and worker well-being. She has examined the impact of these people factors and situational constraints as they affect skill training, job search, teamwork, job performance, and the development of workplace competencies. Her recent projects have focused on adult development and workforce gaining, job search–employment relations, motivation in and of teams, and person determinants of cross-cultural effectiveness. She is director of the Work Science Center and codirector of the Kanfer-Ackerman laboratory, which conducts longitudinal and large-scale laboratory and field collaborative projects on such topics as workforce aging, work adjustment, cognitive fatigue, skill acquisition, adult development and career trajectories, and self-regulated learning. She is a fellow of the Association for Psychological Sciences and the Academy of Management. She has a Ph.D. in psychology from Arizona State University.

Maria Lytell is a senior behavioral scientist and associate director of the Personnel, Training, and Health Program in the Arroyo Center at the RAND Corporation in Arlington, Virginia. She assists in the oversight of a portfolio of studies for the U.S. Army that span five research streams: total workforce management, recruiting and retention, leader development, training readiness and effectiveness, and soldier and family wellness and support. In addition, she leads and contributes to projects that include diversity in the U.S. military, career field classification for U.S. Air Force personnel, and proficiency of U.S. Army enlisted intelligence analysts. She was previously a member of the research staff of the Military Leadership Diversity Commission, a large commission mandated by Congress to recommend how the U.S. military can increase diversity in its top ranks. She also held two internships as a graduate student, one for a private firm designing personnel selection systems and the other for a state-level government office refining civil service job classification series. She has an M.A. and a Ph.D. in industrial-organizational psychology from the University of Illinois at Urbana-Champaign.

Michael S. North is an assistant professor of management and organizations at the New York University Stern School of Business. His research focuses primarily on age, ageism, generational issues, and related management and policy issues. At the Stern School, he is the founding director of the Accommodating Generations in Employment (AGE) Initiative, which conducts research pertaining to the increasingly older and intergenerational workplace and workforce. His work is aimed at identifying strategies for businesses, policy makers, and society to adapt to multigenerational workforce trends. He was recently designated a "rising star" by the Association for Psychological Science. He has a B.A. in psychology from the University of Michigan and a Ph.D. in psychology and social policy from Princeton University.

Julie Anne Schuck (*Study Director*) is a program officer with the Board on Behavioral, Cognitive, and Sensory Sciences at the National Academies of Sciences, Engineering, and Medicine. She has provided analytical, administrative, and editorial support for many studies and workshops and served as a technical writer for many reports. Her projects have covered a wide range of subjects, including law and justice issues; national security; STEM (science, technology, engineering, and mathematic) education; the science of human-system integration, workforce development, and the evaluations of several federal research programs. She has a B.S. in engineering physics from the University of California, San Diego, and an M.S. in education from Cornell University.

Joanne Spetz is the Brenda and Jeffrey L. Kang Presidential chair in Health Care Finance at the Philip R. Lee Institute for Health Policy Studies (IHPS) at the University of California at San Francisco (UCSF). She also is associate director for research at IHPS and a professor in the Department of Family and Community Medicine at UCSF. She is the associate director for research at the Healthforce Center at UCSF and the director of the UCSF Health Workforce Research Center for Long-Term Care. Her fields of specialty are economics of the health care workforce, shortages and supply of registered nurses, organization and quality of the hospital industry, effects of health information technology, effects of medical marijuana policy on youth substance use, and the workforce involved in treatment of substance use disorders. She studied at the Massachusetts Institute of Technology and has a Ph.D. in economics from Stanford University.

Mo Wang is the Lanzillotti-McKethan eminent scholar chair in the Warrington College of Business, the chair of the Management Department, and director of the Human Resource Research Center, all at the University of Florida. He specializes in research on retirement and older worker employment, occupational health psychology, and advanced quantitative methodologies. The Human Resources Research Center contributes to both the science and the profession of human resource management by supporting educational programs and research that focus on factors that affect human performance in work settings. He was the editor of the *Oxford Handbook of Retirement* and currently serves as the editor-in-chief for *Work, Aging and Retirement*. Previously, he served as president of the Society for Occupational Health Psychology and director for the Science of Organizations Program at the National Science Foundation. He has a Ph.D. in industrial-organizational psychology and developmental psychology from Bowling Green State University.